VERTEBRATA
MOLLUSCA
RADIATA
ARTICULATA

BIRDS
MAMMALS
FISHES
REPTILES
CRUSTACEA
INSECTS
WORMS
ECHINODERMS
POLYPI
ACALEPHÆ
ACEPHALA
GASTEROPODS
CEPHALOPODS

SWIMMERS
WADERS
CLIMBERS
PERCHERS
MAN
CARNIVORA
HERBIVORA
WHALES
MARSUPIALS
CYCLOIDS
CTENOIDS
PLACOIDS
CANOIDS
RHIZODONTS
SERPENTS
LIZARDS
TURTLES
FROGS
MALACOSTRACANS
ENTOMOSTRACANS
TRILOBITES
MANDUCATA
SUCTORIA
APTERA
TUBULIBRANCHIATES
DORSIBRANCHIATES
ABRANCHIATES
HOLOTHURIANS
ECHINI
ASTEROIDS
CRINOIDS
ACTINOIDS
RHIZOPODS
HYDROIDS
CTENAPHORI
DISCOPHORI
SIPHONOPHORI
BRYOZOA
BRACHIOPODS
LAMELLIBRANCHIATES
BRANCHIFERS
PULMONATES
PTEROPODS
NAUTILI
AMMONITES
CUTTLE-FISHES

I
II
III
IV

REIGN OF FISHES
REIGN OF REPTILES
REIGN OF MAMMALS
REIGN OF MAN

IV. Modern Age.		
III. Tertiary Age.		Upper Tertiary Formation.
		Lower Tertiary "
II. Secondary Age.		Cretaceous "
		Oölitic "
		Trias "
		Carboniferous "
I Palæozoic Age.		Devonian "
		Upper Silurian "
		Lower Silurian "
Metamorphic Rocks.		

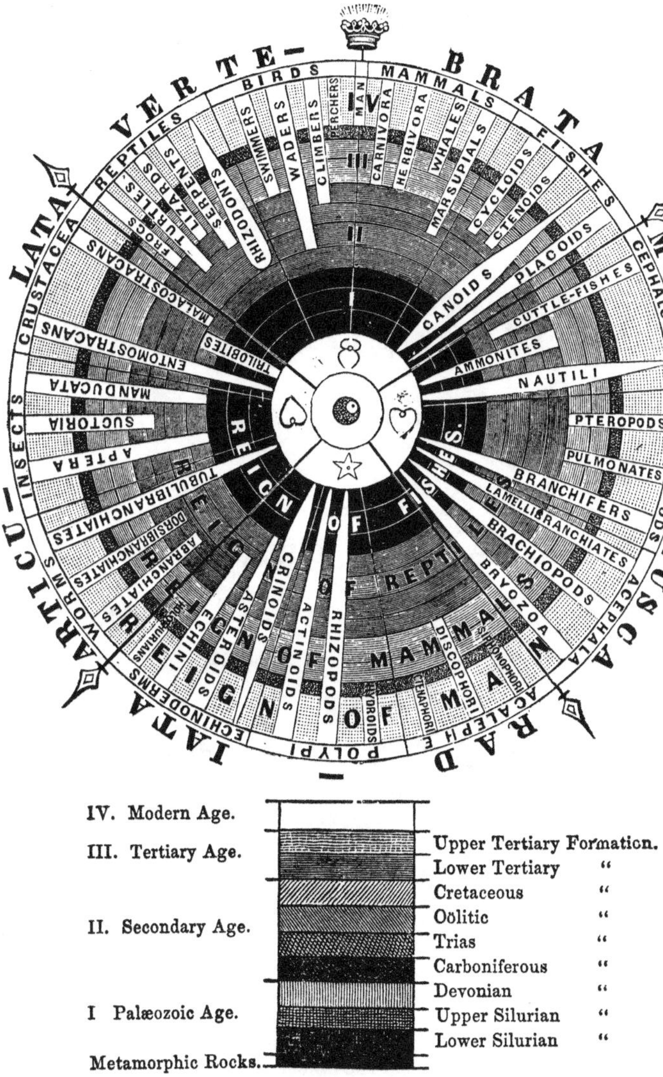

CRUST OF THE EARTH AS RELATED TO ZOÖLOGY.

PRINCIPLES OF ZOÖLOGY:

TOUCHING

THE STRUCTURE, DEVELOPMENT, DISTRIBUTION,
AND NATURAL ARRANGEMENT

OF THE

RACES OF ANIMALS, LIVING AND EXTINCT

WITH NUMEROUS ILLUSTRATIONS.

PART I.

COMPARATIVE PHYSIOLOGY.

FOR THE USE OF SCHOOLS AND COLLEGES.

BY

LOUIS AGASSIZ AND A. A. GOULD.

REVISED EDITION.

BOSTON:
GOULD AND LINCOLN,
59 WASHINGTON STREET.
NEW YORK: SHELDON, BLAKEMAN & CO.
CINCINNATI: GEO. S. BLANCHARD.
1857.

G. C. Rand & Co., Printers, Cornhill.

PREFACE.

The design of this work is to furnish an epitome of the leading principles of the science of Zoölogy, as deduced from the present state of knowledge, so illustrated as to be intelligible to the beginner. No similar treatise now exists in this country, and, indeed, some of the topics have not been touched upon in the English language, unless in a strictly technical form, and in scattered articles. On this account, some of the chapters, like those on Embryology and Metamorphosis, may, at first, seem too abstruse for scholars in our common schools. This may be the case, until teachers shall have made themselves somewhat familiar with subjects comparatively new to them. But so essential have these subjects now become to a correct interpretation of philosophical zoölogy, that the study of them will hereafter be indispensable. They furnish a key to many phenomena which have been heretofore locked in mystery.

Being intended for American students, the illustrations have been drawn, as far as possible, from American objects : some of them are presented merely as ideal outlines, which convey a more definite idea than accurate sketches from nature ; others have been left imperfect, except as to the parts especially in question ; a large proportion of them, however, are accurate portraits from original drawings. Popular names have been employed as far as possible, and to the scientific names an English termination has generally been given ; but the technical terms have been added, in brackets, whenever misunderstanding was apprehended. Definitions of those least likely to be understood, may be found in the Index.

The principles of Zoölogy developed by Professor Agassiz in his published works have been generally adopted in this, and the results of many new researches have been added.

The authors gratefully acknowledge the aid they have received, in preparing the illustrations and working out the details, from Mr.

E. Desor for many years an associate of Professor Agassiz, from Count Pourtalès and E. C. Cabot, Esq., and also from Professor Asa Gray, by valuable suggestions in the revision of the letter-press.

The first part is devoted to Comparative Anatomy, Physiology, and Embryology, as the basis of Classification, and also to the illustration of the geographical distribution and the geological succession of Animals; the second to Systematic Zoölogy, in which the principles of Classification will be applied, and the principal groups of animals will be briefly characterized.

Should our aim be attained, this work will produce more enlarged ideas of man's relations to Nature, and more exalted conceptions of the Plan of Creation and its Great Author.

BOSTON, *June* 1, 1848.

PREFACE TO THE REVISED EDITION.

IN revising the present work, the authors have endeavored to render more precise those passages which admitted of too broad a signification or of a double interpretation; and to correct such errors as had arisen from inadvertence, or such as the rapid progress of Science has disclosed. They are indebted for many suggestions on these points to several distinguished teachers who have used the work as a text book, and more especially to Professor Wyman, of Harvard University. Several entirely new paragraphs have also been added.

A list of some of the principal authors who have made original researches, or of treatises which enter more into detail than was admissible in an elementary work, has been given at the close of the volume, for the use of those who would pursue the subject of Zoölogy in a more extended manner.

The work having thus been revised and enlarged, the authors submit it to the public with increased confidence in its accuracy and usefulness.

BOSTON, *February* 1, 1851.

TABLE OF CONTENTS.

SECTION II.

CHAPTER FOURTH.

CHAPTER FIFTH.

SECTION I.

SECTION II.

CHAPTER SIXTH.

SECTION I.

CHAPTER SEVENTH.

CHAPTER EIGHTH.

CHAPTER NINTH.

CHAPTER TENTH.

SECTION I.

SECTION II.

SECTION III.

CHAPTER ELEVENTH.

SECTION I.

SECTION II.

SECTION III.

CHAPTER TWELFTH.

CHAPTER THIRTEENTH.

SECTION I.

SECTION II.

SECTION III.

CHAPTER FOURTEENTH.

SECTION I.

SECTION II.

EXPLANATION OF THE FIGURES.

FRONTISPIECE.—The diagram opposite the title page is intended to present, at one view, the distribution of the principal types of animals, and the order of their successive appearance in the layers of the earth's crust. The four Ages of Nature, mentioned at page 221, are represented by four zones, of different shades, each of which is subdivided by circles, indicating the number of formations of which they are composed. The whole disk is divided by radiating lines into four segments, to include the four great departments of the Animal Kingdom; the Vertebrates, with Man at their head, are placed in the upper compartment, the Articulates at the left, the Mollusks at the right, and the Radiates below, as being the lowest in rank. Each of these compartments is again subdivided to include the different classes belonging to it, which are named at the outer circle. At the centre is placed a figure to represent the primitive egg, with its germinative vesicle and germinative dot, (278,) indicative of the universal origin of all animals, and the epoch of life when all are apparently alike, (275, 276.) Surrounding this, at the point from which each department radiates, are placed the symbols of the several departments, as explained on page 155. The zones are traversed by rays which represent the principal types of animals, and their origin and termination indicates the age at which they first appeared or disappeared, all those which reach the circumference being still in existence. The width of the ray indicates the greater or less prevalence of the type at different geological ages. Thus, in the class of Crustaceans, the Trilobites appear to commence in the earliest strata, and to disappear with the carboniferous formation. The Ammonites also appeared in the Silurian formation, and did not become extinct before the deposition of the Cretaceous rocks. The Belemnites appear in the lower Oölitic beds; many forms commence in the Tertiary; a great number of types make their appearance only in the Modern age; while only a few have continued from the Silurian, through every period to the present. Thus, the Crinoids were very numerous in the Primary Age, and are but slightly developed in the Tertiary and Modern Age. It is seen, at a glance, that the Animal Kingdom is much more diversified in the later than in the earlier Ages.

Below the circle is a section, intended to show more distinctly the relative position of the ten principal formations of stratified rocks (461) composing the four great geological ages; the numerals corresponding to those on the ray leading to Man, in the circular figure. See also figure 154.

The Chart of Zoölogical Regions, page 195, is intended to show the limits of the several Faunas of the American Continent, corresponding to the climatal regions. And as the higher regions of the mountains correspond in temperature to the climate of higher latitudes, it will be seen that the northern temperate fauna extends, along the mountains of Mexico and Central America, much farther towards the Equator than it does on the lower levels. In the same manner, the southern warm fauna extends northward, along the Andes.

Fig.

1. Simple cell, magnified, as seen in the house-leek.
2. Cells when altered by pressure upon each other; from the pith of elder.
3. Nucleated cells, (*a*,) magnified; *b*, nucleolated cells.
4. Cartilaginous tissue from a horse, magnified 120 diameters.
5. Osseous tissue from a horse, magnified 450 diameters.
6. Nervous fibres, showing the loops as they terminate in the skin of a frog.
7. Gray substance of the brain, magnified.
8. Head of an embryo fish, to show its cellular structure throughout.
9. Diagram, to show the nervous system of the Vertebrates, as found in a monkey.
10. Diagram of the nervous system of the Articulates, as seen in a lobster.
11. Diagram of the nervous system of the Mollusks, as found in *Natica heros*.
12. Diagram of the nervous system of the Radiates, as found in Scutella, (*Echinarachnius parma.*)
13. Section of the eye. *a*, optic nerve; *b*, sclerotic coat; *c*, choroid coat; *d*, retina; *e*, crystalline lens; *f*, cornea; *g*, iris; *h*, vitreous body; *i*, chamber, divided by the iris.
14. Diagram, showing the effect of the eye on rays of light.
15. Position of the eye of the snail.
16. Eyes (*ocelli*) of the spider.
17. Eye-spots of a star-fish, (*Echinaster sanguinolentus.*)
18. Compound eyes, showing the arrangement of the facettes, and their connection with the optic nerve, as seen in a crab's eye.
19. Diagram of the human ear, to show the different chambers, canals, and bones.
20. Tympanum and small bones of the ear, twice the natural size; *c*, tympanum; *m*, malleus; *n*, incus; *o*, orbiculare: *s*, stapes.
21. Section of the brain of a crow, showing the origin of the nerves of the special senses.
22. Diagram of the larynx, in man.
23. Larynx of the merganser, (*Mergus merganser.*)
24. Nests of Ploceus Philippinus, male and female.
25. Distribution of nerves to the muscular fibres.
26. Test, or crust-like covering of an Echinoderm, (*Cidaris.*)

Fig.

77. A polyp, (*Tubularia indivisa;*) *m*, mouth; *o*, ovaries; *p*, tentacles
78. Blood disks in man, magnified.
79. " " in birds, "
80. " " in reptiles, "
81. " " in fishes, "
82. Portion of a vein opened, to show the valves.
83. Network of capillary vessels.
84. Dorsal vessel of an insect, with its valves.
85. Cavities of the heart of mammals and birds.
86. " " " of a reptile.
87. " " " of a fish.
88. Heart and bloodvessels of a gasteropod mollusk, (*Natica.*)
89. Tracheæ, or air tubes of an insect; *s*, stigmata; *t*, trachea.
90. Relative position of the heart and lungs in man.
91. Respiratory organs of a naked mollusk, (*Polycera illuminata.*)
92. Respiratory organs (gills) of a fish.
93. Vesicles and canals of the salivary glands.
94. Section of the skin, magnified, to show the sweat glands; *a*, the cutis; *b*, blood-layer; *c*, epidermis; *g*, gland imbedded in the fat-layer, (*f*.)
95. Egg of a skate-fish, (*Myliobatis.*)
96. Egg of hydra.
97. Egg of snow-flea, (*Podurella.*)
98. Section of an ovarian egg; *d*, germinative dot; *g*, germinative vesicle; *s*, shell membrane; *v*, vitelline membrane.
99. Egg cases of Pyrula.
100. Monoculus bearing its eggs, *a a*.
101. Section of a bird's egg; *a*, albumen; *c*, chalaza; *e*, embryo; *s*, shell; *y*, yolk.
102. Cell-layer of the germ.
103. Separation of the cell-layer into three layers; *s*, serous or nervous layer; *m*, mucous or vegetative layer; *v*, vascular or blood layer.
104. Embryo of a crab, showing its incipient rings.
105. Embryo of a vertebrate, showing the dorsal furrow.

106-8. Sections of the embryo, showing the formation of the dorsal canal.

109. Section, showing the position of the embryo of a vertebrate, in relation to the yolk.
110. Section, showing the same in an articulate, (*Podurella.*)

111-22. Sections, showing the successive stages of development of the embryo of the white-fish, magnified.

123. Young white-fish just escaped from the egg, with the yolk not yet fully taken in.

124, 125. Sections of the embryo of a bird, showing the formation of the allantois; *e*, embryo; *x x*, membrane rising to form the amnios; *a*, allantois; *y*, yolk.

126. The same more fully developed. The allantois (*a*) is further de-

FIG.

veloped, and bent upwards. The upper part of the yolk (*d d*) is nearly separated from the yolk sphere, and is to become the intestine. The heart (*h*) is already distinct, and connected by threads with the blood-layer of the body.

127. Section of the egg of a mammal; *v*, the thick vitelline membrane, or chorion; *y*, yolk; *s*, germinative dot; *g*, germinative vesicle.
128. The same, showing the empty space (*k*) between the vitelline sphere and chorion.
129. Shows the first indications of the germ already divided in two layers, the serous layer, (*s*,) and the mucous layer, (*m*.)
130. The mucous layer (*m*) expands over nearly half of the yolk, and becomes covered with many little fringes.
131. The embryo (*e*) is seen surrounded by the amnios, (*b*,) and covered by a large allantois, (*a*;) *p e*, fringes of the chorion; *p m*, fringes of the matrix.
132. Hydra, showing its reproduction by buds.
133. Vorticella, showing its reproduction by division.
134. Polyps, showing the same.
135. A chain of Salpæ.
136. An individual salpa; *m*, the mouth; *a*, embryos.
137. Cercaria, or early form of the Distoma.
138. Distoma, with its two suckers.
139. Nurse of the Cercaria.
140. The same, magnified, showing the included young.
141. Grand nurses of the Cercaria, enclosing the young nurses.
142. Stages of development of a jelly-fish, (*Medusa*;) *a*, the embryo in its first stage, much magnified; *b*, summit, showing the mouth; *c*, *f*, *g*, tentacles shooting forth; *e*, embryo adhering, and forming a pedicle; *h*, *i*, separation into segments; *d*, a segment become free; *k*, form of the adult.
143. Portion of a plant-like polyp, (*Campanularia*) *a*, the cup which bears tentacles; *b*, the female cup, containing eggs; *c*, the cups in which the young are nursed, and from which they issue.
144. Young of the same, with its ciliated margin, magnified.
145. Eye of the perch, containing parasitic worms, (*Distoma*.)
146. One of the worms magnified.
147. Transformations of the canker-worm, (*Geometra vernalis*;) *a*, the canker worm; *b*, its chrysalis; *c*, female moth; *d*, male moth.
148. Metamorphoses of the duck-barnacle, (*Anatifa*;) *a*, eggs, magnified; *b*, the animal as it escapes from the egg; *c*, the stem and eye appearing, and the shell enclosing them; *d*, animal removed from the shell, and further magnified; *e*, *f*, the mature barnacle, affixed.
149. Metamorphoses of a star-fish, (*Echinaster sanguinolentus*,) showing the changes of the yolk, (*e*;) the formation of the pedicle, (*p*;) and the gradual change into the pentagonal and rayed form.

FIG.

150. Comatula, a West India species, in its early stage, with its stem.
151. The same detached, and swimming free.
152. Longitudinal section of the sturgeon, to show its cartilaginous vertebral column.
153. Amphioxus, natural size, showing its imperfect organization.
154. Section of the earth's crust, to show the relative positions of the rocks composing it; *E*, plutonic or massive rocks; *M*, metamorphic rocks; *T*, trap rocks; *L*, lava. 1. Lower Silurian formation; 2. Upper Silurian; 3. Devonian; 4. Carboniferous; 5. Trias, or Saliferous; 6. Oölitic; 7. Cretaceous; 8. Lower Tertiary or Eocene; 9. Upper Tertiary, or Miocene, and Pleiocene; 10. Drift.
155. Fossils of the Palæozoic age; *a*, Lingula prima; *b*, Leptæna alternata; *c*, Euomphalus hemisphericus; *d*, Trocholites ammonius; *e*, Avicula decussata; *f*, Bucania expansa; *g*, Orthoceras fusiforme; *i*, Cyathocrinus ornatissimus, Hall; *j*, Cariocrinus ornatus, Say; *k*, Melocrinus amphora, Goldf.; *l*, Columnaria alveolata; *m*, Cyathophyllum quadrigeminum, Goldf.; *n*, *o*, Caninia flexuosa; *p*, Chætetes lycoperdon.
156. Articulata of the Palæozoic age; *a*, Harpes; *b*, Arges; *c*, Brontes; *d*, Platynotus; *e*, Eurypterus remipes.
157. Fishes of the Palæozoic age; *a*, Pterichthys; *b*, Coccosteus; *c*, Dipterus; *d*, palatal bone of a shark; *e*, spine of a shark.
158. Representations of the tracks of supposed birds and reptiles in the sandstone rocks.
159. Supposed outlines of Ichthyosaurus, (*a*,) and Plesiosaurus, (*b*.)
160. Supposed outline of Pterodactyle.
161. Shells of the Secondary age; *a*, Terebratula; *b*, Goniomya; *c*, Trigonia; *d*, Ammonite.
162. Supposed outline of the cuttle-fish, (*a*,) furnishing the Belemnite.
163. Radiata from the Secondary age; *a*, Lobophyllia flabellum; *b*, Lithodendron pseudostylina; *c*, Pentacrinus briareus; *d*, Pterocoma pinnata; *e*, Cidaris; *f*, Dysaster; *g*, Nucleolites.
164. Shells of the Cretaceous formation; *a*, Ammonites; *b*, Crioceras; *c*, Scaphites; *d*, Ancyloceras; *e*, Hamites; *f*, Baculites; *g*, Turrilites.
165. Shells of the Cretaceous formation; *a*, Magas; *b*, Inoceramus; *c*, Hippurites; *d*, Spondylus; *e*, Pleurotomaria.
166. Radiata from the Cretaceous formation; *a*, Diploctenium cordatum; *b*, Marsupites; *d*, Galerites; *c*, Salenia; *e*, Micraster coranguinum.
167 Nummulite.
168. Supposed outline of Paleotherium.
169. Supposed outline of Anoplotherium.
'70 Skeleton of the Mastodon, in the cabinet of Dr. J. C. Warren.

INTRODUCTION.

Every art and science has a language of technical terms peculiar to itself. With those terms every student must make himself familiarly acquainted at the outset; and, first of all, he will desire to know the names of the objects about which he is to be engaged.

The names of objects in Natural History are double; that is to say, they are composed of two terms. Thus, we speak of the white-bear, the black-bear, the hen-hawk, the sparrow-hawk; or, in strictly scientific terms, we have *Felis leo*, the lion, *Felis tigris*, the tiger, *Felis catus,* the cat, *Canis lupus*, the wolf, *Canis vulpes,* the fox, *Canis familiaris*, the dog, &c. They are always in the Latin form, and consequently the adjective name is placed last. The first is called the *generic* name; the second is called the *trivial,* or *specific* name.

These two terms are inseparably associated in every object of which we treat. It is very important, therefore, to have a clear idea of what is meant by the terms *genus* and *species;* and although the most common of all others, they are not the easiest to be clearly understood. The Genus is founded upon some of the minor peculiarities of anatomica. structure, such as the number, disposition, or proportions of the teeth, claws, fins, &c., and usually includes several kinds. Thus, the lion, tiger, leopard, cat, &c., agree in the structure of their feet, claws, and teeth, and they belong to the genus Felis; while the dog, fox, jackal, wolf, &c., have another and a different peculiarity of the feet, claws, and teeth, and are arranged in the genus Canis.

The Species is founded upon less important distinctions, such as color, size, proportions, sculpture, &c. Thus we have different kinds, or species, of duck, different species of squirrel, different species of monkey, &c., varying from each other in some trivial circumstance, while those of each group agree in all their general structure. The specific name is the lowest term to which we descend, if we except certain peculiarities, generally induced by some modification of native habits, such as are seen in domestic animals. These are called *varieties*, and seldom endure beyond the causes which occasion them.

Several genera which have certain traits in common are combined to form a *family*. Thus, the alewives, herrings, shad, &c., form a family called Clupeidæ; the crows, blackbirds, jays, &c., form the family Corvidæ. Families are combined to form *orders*, and orders form *classes*, and finally, classes are combined to form the four primary divisions, or *departments*, of the Animal Kingdom.

For each of these groups, whether larger or smaller, we involuntarily picture in our minds an image, made up of the traits which characterize the group. This ideal image is called a TYPE, a term which there will be frequent occasion to employ in our general remarks on the Animal Kingdom. This image may correspond to some one member of the group; but it is rare that any one species embodies all our ideas of the class, family, or genus to which it belongs.

Thus, we have a general idea of a bird; but this idea does not correspond to any particular bird, or any particular character of a bird. It is not precisely an ostrich, an owl, a hen, or a sparrow; it is not because it has wings, or feathers, or two legs; or because it has the power of flight, or builds nests. Any, or all, of these characters would not fully represent our idea of a bird; and yet every one has a distinct ideal notion of a bird, a fish, a quadruped, &c. It is common, however, to speak of the animal which embodies most fully the characters of a group, as the type of that group. Thus we might, perhaps, regard an eagle as the type of a bird, the duck as the type of a swimming-bird, and the mallard as the type of a duck, and so on.

As we must necessarily make frequent allusions to animals, with reference to their systematic arrangement, it seems requisite to give a sketch of their classification in as popular terms as may be, before entering fully upon that subject, and with particular reference to the diagram fronting the title-page.

The Animal Kingdom consists of four great divisions, which we call DEPARTMENTS, namely:

I. The department of Vertebrates.
II. The department of Articulates.
III. The department of Mollusks.
IV. The department of Radiates.

I. The department of VERTEBRATES includes all animals which have an internal skeleton, with a back-bone for its axis. It is divided into four classes:

1. Mammals, (animals which nurse their young.)
2. Birds.

3. Reptiles.
4. Fishes.

The class of Mammals is subdivided into three orders:

a. Beasts of prey, (*Carnivora.*)
b. Those which feed on vegetables, (*Herbivora.*)
c. Animals of the whale kind, (*Cetaceans.*)

The class of Birds is divided into four orders, namely,

a. Perching Birds, (*Insessores.*)
b. Climbers, (*Scansores.*)
c. Waders, (*Grallatores.*)
d. Swimmers, (*Natatores.*)

The class of Reptiles is divided into five orders:

a. Large reptiles with hollow teeth, most of which are now extinct, (*Rhizodonts.*)
b. Lizards, (*Lacertians.*)
c. Snakes, (*Ophidians.*)
d. Turtles, (*Chelonians.*)
e. Frogs and Salamanders, (*Batrachians.*)

The class of Fishes is divided into four orders:

a. Those with enamelled scales, like the gar-pike, (*Ganoids,*) fig. 157, *c.*
b. Those with the skin like shagreen, as the sharks and skates, (*Placoids.*)
c. Those which have the edge of the scales toothed, and usually with some bony rays to the fins, as the perch, (*Ctenoids.*)

d. Those whose scales are entire, and whose fin rays are soft, like the salmon, (*Cycloids.*)

II. Department of ARTICULATES. Animals whose body is composed of rings or joints. It embraces three classes:

1. Insects.
2. Crustaceans, like the crab, lobster, &c.
3. Worms.

The class of INSECTS includes three orders:

a. Those with a trunk for sucking fluids, like the butterfly, (*Suctoria,*) fig. 62–64.

b. Those which have jaws for dividing their food, (*Manducata,*) fig. 60.

c. Those destitute of wings, like spiders, fleas, millipedes, &c., (*Aptera.*)

The class CRUSTACEANS may be divided as follows:

a. Those furnished with a shield, like the crab and lobster, (*Malacostraca.*)

b. Such as are not thus protected, (*Entomostraca.*)

c. An extinct race, intermediate between these two, (*Trilobites,*) fig. 156.

The class of WORMS comprises three orders:

a. Those which have thread-like gills about the head, (*Tubulibranchiates.*)

b. Those whose gills are placed along the sides, (*Dorsibranchiates.*)

c. Those who have no exterior gills, like the earth-worm (*Abranchiates,*) and also the Intestinal Worms.

III. The department of MOLLUSKS is divided into three classes, namely:

1. Those which have arms about the mouth, like the cuttle-fish, (*Cephalopods*,) fig. 47.
2. Those which creep on a flattened disk or foot, like snails, (*Gasteropods*,) fig. 88.
3. Those which have no distinct head, and are inclosed in a bivalve shell, like the clams, (*Acephals.*)

The CEPHALOPODS may be divided into

a. The cuttle-fishes, properly so called, (*Teuthideans*,) fig. 47.

b. Those having a shell, divided by sinuous partitions into numerous chambers, (*Ammonites*,) fig. 164.

c. Those having a chambered shell with simple partitions, (*Nautilus.*)

The GASTEROPODS contain four orders:

a. The land snails which breathe air, (*Pulmonates.*)

b. The aquatic snails which breathe water, (*Branchifers*,) fig. 88.

c. Those which have wing-like appendages about the head, for swimming, (*Pteropods.*)

d. A still lower form allied to the Polyps by their general appearance, (*Rhizopods* or *Foraminifera.*)

The class of ACEPHALS contains three orders:

a. Those having shells of two valves, (bivalves,) like the clam and oyster, (*Lamellibranchiates.*)

b. Those having two unequal valves, and furnished with peculiar arms, (*Brachiopods.*)

c. Mollusks living in chains or clusters, like the Salpa, fig. 135; or upon plant-like stems, like Flustra, (*Bryozoa.*)

IV. The department of RADIATES is divided into three classes:

1. Sea-urchins, bearing spines upon the surface, (*Echinoderms,*) figs. 12, 26.
2. Jelly-fishes, (*Acalephs,*) fig. 31.
3. Polyps, fixed like plants, and with a series of flexible arms around the mouth, figs. 48, 77, 143.

The ECHINODERMS are divided into four orders:

a. Sea-slugs, like biche-le-mar, (*Holothurians.*)

b. Sea-urchins, (*Echini,*) fig. 26.

c. Free star-fishes, (*Asteridæ,*) fig. 17.

d. Star-fishes mostly attached by a stem, (*Crinoids,*) figs. 150, 151.

The ACALEPHS include the following orders:

a. Those furnished with vibrating hairs, by which they move, (*Ctenophoræ.*)

b. The Medusæ, or common jelly-fishes, (*Discophoræ,*) figs. 31, 142.

c. Those provided with aerial vesicles, (*Siphonophoræ.*)

The class of POLYPS includes two orders.

a. The so-called fresh-water polyps, and similar marine forms, with lobed tentacles, (*Hydroïds,*) fig. 143.

b. Common polyps, like the sea-anemone and coral-polyp, (*Actinoids,*) fig. 48.

In addition to these, there are numberless kinds of microscopic animalcules, commonly united under the name of infusory animals, (Infusoria,) from their being found specially abundant in water infused with vegetable matter. These minute beings do not, however, constitute a natural group in the Animal Kingdom. Indeed, a great many that were formerly supposed to be animals are now found to be vegetables. Others are ascertained to be crustaceans, mollusks, worms of microscopic size, or the earliest stages of development of larger species. In general, however, they are exceedingly minute, and exhibit the simplest forms of animal life, and are now grouped together, under the title of Protozoa. But, as they are still very imperfectly understood, notwithstanding the beautiful researches already published on this subject, and as many of them are likely to be finally distributed among vegetables, and the legitimate classes in the Animal Kingdom to which they belong, we have not assigned any special place for them.

PHYSIOLOGICAL ZOÖLOGY.

CHAPTER FIRST.

THE SPHERE AND FUNDAMENTAL PRINCIPLES OF ZOÖLOGY.

1. ZOÖLOGY is that department of Natural History which relates to animals.

2. To enumerate and name the animals which are found on the globe, to describe their forms, and investigate their habits and modes of life, are the principal, but by no means the only objects of this science. Animals are worthy of our regard, not merely when considered as to the variety and elegance of their forms, or their adaptation to the supply of our wants; but the Animal Kingdom, as a whole, has a still higher signification. It is the exhibition of the divine thought, as carried out in one department of that grand whole which we call Nature; and considered as such, it teaches us most important lessons.

3. Man, in virtue of his twofold constitution, the spiritual and the material, is qualified to comprehend Nature.

Being made in the spiritual image of God, he is competent to rise to the conception of His plan and purpose in the works of Creation. Having also a material body, like that of other animals, he is also in a condition to understand the mechanism of organs, and to appreciate the necessities of matter, as well as the influence which it exerts over the intellectual element throughout the domain of Nature.

4. The spirit and preparation we bring to the study of Nature, is a matter of no little consequence. When we would study with profit a work of literature, we first endeavor to make ourselves acquainted with the genius of the author; and in order to know what end he had in view, we must have regard to his previous labors, and to the circumstances under which the work was executed. Without this, although we may perhaps enjoy its perfection as a whole, and admire the beauty of its details, yet the spirit which pervades it will escape us, and many passages may even remain unintelligible.

5. So, in the study of Nature, we may be astonished at the infinite variety of her products; we may even study some portion of her works with enthusiasm, and nevertheless remain strangers to the spirit of the whole, ignorant of the plan on which it is based, and fail to acquire a proper conception of the varied affinities which combine beings together, so as to make of them that vast picture in which each animal, each plant, each group, each class, has its place, and from which nothing could be removed without destroying the proper meaning of the whole.

6. Besides the beings which inhabit the earth at the present time, this picture also embraces the extinct races which are now known to us by their fossil remains only. And these are of the greatest importance, since they furnish us with the means of ascertaining the changes and modifications which the Animal Kingdom has undergone in the suc-

cessive creations, since the first appearance of living beings.

7. It is but a short time since it was not difficult for a man to possess himself of the whole domain of positive knowledge in Zoölogy. A century ago, the number of known animals did not exceed 8000; that is to say, from the whole Animal Kingdom, fewer species were then known than are now contained in many private collections of certain families of insects merely. At the present day, the number of living species which have been satisfactorily made out and described, is more than 50,000.* The fossils already described exceed 6000 species; and if we

---

* The number of vertebrate animals may be estimated at 20,000. About 1500 species of mammals are pretty precisely known, and the number may probably be carried to about 2000.

The number of Birds well known is 4 or 5000 species, and the probable number is 6000.

The Reptiles number about the same as the Mammals, 1500 described species, and they will probably reach the number of 2000.

The Fishes are more numerous: there are from 5 to 6000 species in the museums of Europe, and the number may probably amount to 8 or 10,000.

The number of Mollusks already in collections probably reaches 8 or 10,000. There are collections of marine shells, bivalve and univalve, which amount to 5 or 6000; and collections of land and fluviatile shells, which count as many as 2000. The total number of mollusks would, therefore, probably exceed 1[illegible]0 species.

Among the articulated animals it is difficult to estimate the number of species. There are collections of coleopterous insects which number 20 to 25,000 species; and it is quite probable, that by uniting the principal collections of insects, 60 or 80,000 species might now be counted; for the whole department of articulata, comprising the crustacea, the cirrhipeda, the insects, the red-blooded worms, the intestinal worms, and the infusoria so far as they belong to this department, the number would already amount to 100,000; and we might safely compute the probable number of species actually existing at double that sum.

Add to these about 10,000 for radiata, including echini, star-fishes, medusæ, and polypi, and we have about 250,000 species of living animals; and supposing the number of fossil species only to equal them, we have, at a very moderate computation, half a million of species.

consider that wherever any one stratum of the earth has been well explored, the number of species discovered has not fallen below that of the living species which now inhabit any particular locality of equal extent, and then bear in mind that there is a great number of geological strata, we may anticipate the day when the ascertained fossil species will far exceed the living species.*

8. These numbers, far from discouraging, should, on the contrary, encourage those who study Natural History. Each new species is, in some respects, a radiating point which throws additional light on all around it; so that, as the picture is enlarged, it at the same time becomes more intelligible to those who are competent to seize its prominent traits.

9. To give a detailed account of each and all of these animals, and to show their relations to each other, is the task of the Naturalist. The number and extent of the volumes already published upon the various departments of Natural History show, that only a mere outline of a domain so vast could be fully sketched in an elementary work, and that none but those who make it their special study can be expected to survey its individual parts.

10. Every well-educated person, however, is expected to have a general acquaintance with the great natural phenomena constantly displayed before his eyes. There is a general knowledge of man and the subordinate animals, embracing their structure, races, habits, distribution, mutual relations, &c., which is not only calculated to conduce es-

* In a separate work, entitled "*Nomenclator Zoölogicus*," by L. AGASSIZ, the principles of nomenclature are discussed, and a list of the names of genera and families proposed by authors is given. To this work those are referred who may desire to become more familiar with nomenclature, and to know in detail the genera and families in each class of the Animal Kingdom.

sentially to our happiness, but which it would be quite inexcusable to neglect. This general view of Zoölogy, it is the purpose of this work to afford.

11. A sketch of this nature should render prominent the more general features of animal life, and delineate the arrangement of the species according to their most natural relations and their rank in the scale of being; thus giving a panorama, as it were, of the entire Animal Kingdom. To accomplish this, we are at once involved in the question, What is it that gives an animal precedence in rank?

12. In one sense, all animals are equally perfect. Each species has its definite sphere of action, whether more or less extended, — its own peculiar office in the economy of nature; and a complete adaptation to fulfil all the purposes of its creation, beyond the possibility of improvement. In this sense, every animal is perfect. But there is a wide difference among them, in respect to their organization. In some it is very simple, and very limited in its operation; in others, extremely complicated, and capable of exercising a great variety of functions.

13. In this physiological point of view, an animal may be said to be more perfect in proportion as its relations with the external world are more varied; in other words, the more numerous its functions are. Thus, an animal, like a quadruped, or a bird, which has the five senses fully developed, and which has, moreover, the faculty of readily transporting itself from place to place, is more perfect than a snail, whose senses are very obtuse, and whose motion is very sluggish.

14. In like manner, each of the organs, when separately considered, is found to have every degree of complication, and, consequently, every degree of nicety in the performance of its function. Thus, the eye-spots of the star-fish and jelly-fish are probably endowed with merely the fac

ulty of perceiving light, without the power of distinguishing objects. The keen eye of the bird, on the contrary, discerns minute objects at a great distance, and when compared with the eye of a fly, is found to be not only more perfect, but constructed on an entirely different plan. It is the same with every other organ.

15. We understand the faculties of animals, and appreciate their value, just in proportion as we become acquainted with the instruments which execute them. The study of the functions or uses of organs, therefore, requires an examination of their structure; they must never be disjoined, and must precede the systematic distribution of animals into classes, families, genera, and species.

16. In this general view of organization, we must ever bear in mind the necessity of carefully distinguishing between *affinities* and *analogies*, a fundamental principle recognized even by Aristotle, the founder of scientific Zoölogy. *Affinity* or *homology* is the relation between organs or parts of the body which are constructed on the same plan, however much they vary in form, or even serve for very different uses. *Analogy*, on the contrary, indicates the similarity of purposes or functions performed by organs of different structure.

17. Thus, there is an analogy between the wing of a bird and that of a butterfly, since both of them serve for flight. But there is no affinity between them, since, as we shall hereafter see, they differ totally in their anatomical relations. On the other hand, there is an affinity between the bird's wing and the hand of a monkey; since, although they serve for different purposes, the one for flight, and the other for climbing, they are both constructed on the same plan. Accordingly, the bird is more nearly allied to the monkey than to the butterfly, though they both have in common the faculty of flight. Affinities, and not analogies, therefore, must guide us in the arrangement of animals.

18. Our investigations should not be limited to adult animals, but should also include the changes which they undergo during the whole course of their development. Otherwise, we shall be liable to exaggerate the importance of certain peculiarities of structure which have a predominant character in the full-grown animal, but which are shaded off, and vanish, as we revert to the earlier periods of life.

19. Thus, for example, by regarding only adult individuals, we might be induced to divide all animals into two groups, according to their mode of respiration; uniting, on the one hand, all those which breathe by gills, and, on the other, those which breathe by lungs. But this distinction loses its importance, when we consider that various animals, for example, frogs, which respire by lungs in the adult state, have only gills when young. It is thence evident that the respiratory organs cannot be taken as a satisfactory basis of our fundamental classification. They are, as we shall see, subordinate to a more important system, namely, the nervous system.

20. Again, we have a means of appreciating the relative grade of animals by the comparative study of their development. It is evident that the caterpillar, in becoming a butterfly, passes from a lower to a higher state. Clearly, therefore, animals resembling the caterpillar, the worms, for instance, must occupy a lower rank than those approaching the butterfly, like most insects. There is no animal which does not undergo a series of changes similar to those of the caterpillar or the chicken; only, in many of them, the most important ones occur before birth, during what is called the embryonic period.

21. The life of the chicken has not just commenced when it issues from the egg; for if we break the egg some days previous to the time of hatching, we find in it a living animal, which, although imperfect, is nevertheless a chicken:

it has been developed from a hen's egg, and we know that, should it continue to live, it would infallibly display all the characteristics of the parent bird. Now, if there existed in Nature an adult bird as imperfectly organized as the chicken on the day, or the day before it was hatched, we should assign to it an inferior rank.

22. In studying the embryonic states of the mollusks or worms, we observe in them points of resemblance to many animals of a lower grade, to which they at length become entirely dissimilar. For example, the myriads of minute aquatic animals embraced under the name of Infusoria, generally very simple in their organization, remind us of the embryonic forms of other animals. We shall have occasion to show that the Infusoria are not to be considered as a distinct class of animals, but that among them are found members of all the lower classes of animals, mollusks, crustaceans, worms, &c.; and many of them are even found to belong to the Vegetable Kingdom.

23. Not less striking are the relations that exist between animals and the regions they inhabit. Every animal has its home. Animals of the cold regions are not the same as those of temperate climates; and these latter, in their turn, differ from those of tropical regions. Certainly, no one will maintain it to be the effect of accident that the monkeys, the most perfect of all brute animals, are found only in hot countries; or that by chance merely the white bear and reindeer inhabit only cold regions.

24. Nor is it by chance that most of the largest animals, of every class, the whales, the aquatic birds, the sea-turtles, the crocodiles, dwell in the water rather than on the land. And while the water affords freedom of motion to the largest, it is also the home of the smallest of living beings, allowing a degree of liberty to their motion, which they could not enjoy elsewhere.

25. Nor are our researches to be limited to the animals now living. There are buried in the crust of the earth the remains of a great number of animals belonging to species which do not exist at the present day. Many of these remains present forms so extraordinary that it is almost impossible to trace their alliance with any animal now living. In general, they bear a striking analogy to the embryonic forms of existing species. For example, the curious fossils known under the name of Trilobites (Fig. 156) have a shape so singular that it might well be doubted to what group of articulated animals they belong. But if we compare them with the embryo crab, we find so remarkable a resemblance that we do not hesitate to refer them to the crustaceans. We shall also see that some of the Fishes of ancient epochs present shapes altogether peculiar to themselves, (Fig. 157,) but resembling, in a striking manner, the embryonic forms of our common fishes. A determination of the successive appearance of animals in the order of *time* is, therefore, of much importance in assisting to decide the relative rank of animals.

26. Besides the distinctions to be derived from the varied structure of organs, there are others less subject to rigid analysis, but no less decisive, to be drawn from the immaterial principle with which every animal is endowed. It is this which determines the constancy of species from generation to generation, and which is the source of all the varied exhibitions of instinct and intelligence which we see displayed, from the simple impulse to receive the food which is brought within their reach, as observed in the polyps, through the higher manifestations, in the cunning fox, the sagacious elephant, the faithful dog, to the exalted intellect of man. which is capable of indefinite expansion.

27. Such are some of the general aspects in which we are to contemplate the animal creation. Two points of

view should never be lost sight of, nor disconnected, namely, the animal in respect to its own organism, and the animal in its relations to creation as a whole. By adopting too exclusively either of these points of view, we are in danger of falling either into gross materialism, or into vague and profitless pantheism. He who beholds in Nature nothing besides organs and their functions, may persuade himself that the animal is merely a combination of chemical and mechanical actions and reactions, and thus becomes a materialist.

28. On the contrary, he who considers only the manifestations of intelligence and of creative will, without taking into account the means by which they are executed, and the physical laws by virtue of which all beings preserve their characteristics, will be very likely to confound the Creator with the creature.

29. It is only as it contemplates, at the same time, matter and mind, that Natural History rises to its true character and dignity, and leads to its worthiest end, by indicating to us, in Creation, the execution of a plan fully matured in the beginning, and undeviatingly pursued; the work of a God infinitely wise, regulating Nature according to immutable laws, which He has himself imposed on her.

CHAPTER SECOND.

GENERAL PROPERTIES OF ORGANIZED BODIES.

SECTION I.

ORGANIZED AND UNORGANIZED BODIES.

30. NATURAL HISTORY, in its broadest sense, embraces the study of all the bodies which compose the crust of the earth, or which are dispersed over its surface.

31. These bodies may be divided into two great groups; inorganic bodies, (minerals and rocks,) and living or organized bodies, (vegetables and animals.) These two groups have nothing in common, save the universal properties of matter, such as weight, extension, &c. They differ at the same time as to their form, their structure, their chemical composition, and their mode of existence.

32. The distinctive characteristic of inorganic bodies is *rest;* the distinctive trait of organized bodies is *independent motion*, LIFE. The rock or the crystal, once formed, never changes from internal causes; its constituent parts or molecules invariably preserve the position which they have once taken in respect to each other. Organized bodies, on the contrary, are continually in action. The sap circulates in

the tree, the blood flows through the animal, and in both there is, besides, the incessant movement of growth, decomposition, and renovation.

33. Their mode of formation is also entirely different. Unorganized bodies are either simple or made up of elements unlike themselves; and when a mineral is enlarged, it is simply by the outward addition of particles constituted like itself. Organized bodies are not formed in this manner. They always, and necessarily, are derived from beings similar to themselves; and once formed, they always increase interstitially, by the successive assimilation of new particles, derived from various sources.

34. Finally, organized bodies are limited in their duration. Animals and plants are constantly losing some of their parts by decomposition during life, which at length cease to be supplied, and they die, after having lived for a longer or shorter period. Inorganic bodies, on the contrary, contain within themselves no principle of destruction; and unless subjected to some foreign influence, a crystal or a rock would never change. The limestone and granite of our mountains remain just as they were formed in ancient geological epochs; while numberless generations of plants and animals have lived and perished upon their surface.

SECTION II.

ELEMENTARY STRUCTURE OF ORGANIZED BODIES.

35. The exercise of the functions of life, which is the essential characteristic of organized bodies, (32,) requires a degree of flexibility of the organs. This is secured by means of a certain quantity of watery fluid, which pene-

trates all parts of the body, and forms one of its principal constituents.

36. All living bodies, without exception, are made up of *tissues* so constructed as to be permeable to liquids. There is no part of the body, no organ, however hard and compact it may appear, which has not this peculiar structure. It exists in the bones of animals, as well as in their flesh and fat; in the wood, however solid, as well as in the bark and flowers of plants. It is to this general structure that the term *organism* is now applied. Hence the collective name of *organized beings*,* which includes both the animal and the vegetable kingdoms.

37. The vegetable tissues and most of the organic structures, when examined by the microscope in their early states of growth, are found to be composed of hollow vesicles or *cells*. The natural form of the cells is that of a sphere or of an ellipsoid, as may be easily seen in many plants; for example, in the tissue of the house-leek, (Fig. 1.) The intervals which sometimes separate them from each other are called *intercellular passages* or *spaces* (*m.*) When the cellules are very numerous, and crowd each other, their outlines become angular, and the intercellular spaces disappear, as seen in figure 2, which represents

m

Fig. 1.

* Formerly, animals and plants were said to be *organized*, because they are furnished with definite parts, called *organs*, which execute particular functions. Thus, animals have a stomach, a heart, lungs, &c.; plants have leaves, petals, stamens, pistils, roots, &c., which are indispensable to the maintenance of life and the perpetuation of the species. Since the discovery of the fundamental identity of structure of animal and vegetable tissues, a common denomination for this uniformity of texture has been justly preferred; and the existence of tissues is now regarded as the basis of organization.

the pith of the elder. They then have the form of a honey-comb; whence they have derived their name of cellules.

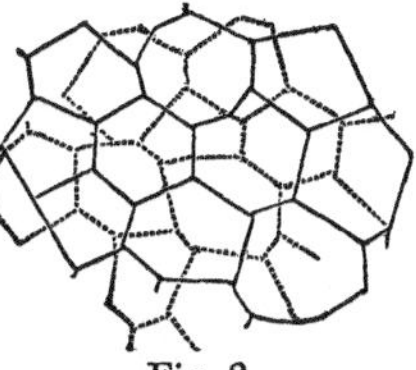
Fig. 2.

38. All the organic tissues, whether animal or vegetable, originate from cells. The cell is to the organized body what the primary form of the crystal is to the secondary, in minerals. As a general fact, it may be stated that animal cells are smaller than vegetable cells; but they alike contain a central dot or vesicle, called *nucleus*. Hence such cells are called *nucleated cells*, (Fig. 3, *a.*) Sometimes the nucleus itself contains a still smaller dot, called *nucleolus*, (*b.*)

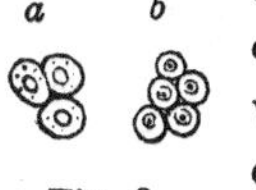

Fig. 3.

39. The elementary structure of vegetables may be ob served in every part of a plant, and its cellular character has been long known. But with the animal tissues there is far greater difficulty. Their variations are so great, and their transformations so diverse, that after the embryonic period it is sometimes impossible, even by the closest examination, to detect their original cellular structure.

40. Several kinds of tissues have been designated in the animal structure; but their differences are not always well marked, and they pass into each other by insensible shades. Their modifications are still the subject of investigation, and we refer only to the most important distinctions.

41. The *areolar tissue* consists of a network of delicate fibres, intricately interwoven so as to leave numberless communicating interstices, filled with fluid. It is interposed in layers of various thickness, between all parts of the body, and frequently accompanied by clusters of fat cells. The fibrous and the serous membranes are mere modifications of this tissue.

42. The *cartilaginous tissue* is composed of nucleated

cells, the intercellular spaces being filled with a more compact substance, called the *hyaline matter*. Figure 4 represents a slip of cartilage from the horse, under a magnifying power of one hundred and twenty diameters.

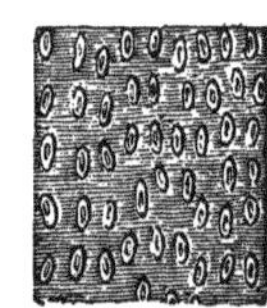
Fig. 4.

43. The *osseous* or *bony tissue* differs from the cartilaginous tissue, in having its meshes filled with salts of lime, instead of hyaline substance, whence its compact and solid appearance. It contains, besides, minute, rounded, or star-like points, improperly called bone-corpuscles, which are found to be cavities or canals, sometimes radiated and branched, as is seen in figure 5, representing a section of a bone of a horse, magnified four hundred times.

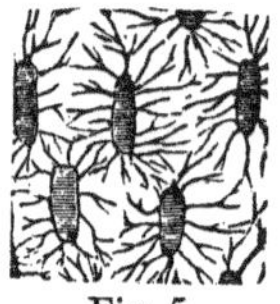
Fig. 5.

44. The *muscular tissue*, which forms the flesh of animals, is composed of bundles of parallel fibres, which possess the peculiar property of contracting or shortening themselves, under the influence of the nerves. In the muscles under the control of the will, the fibres are commonly crossed by very fine lines or wrinkles; but not so in the involuntary muscles. Every one is sufficiently familiar with this tissue, in the form of lean meat.

45. The *nervous tissue* is of different kinds. In the nerves proper, it is composed of very delicate fibres, which return back at their extremities, and form loops, as shown in figure 6, representing nervous threads as they terminate in the skin of a frog. The same fibrous structure is found in the white portion of the brain. But the gray substance of the brain is composed of very minute granulations, interspersed with clusters of larger cells, as seen in figure 7.

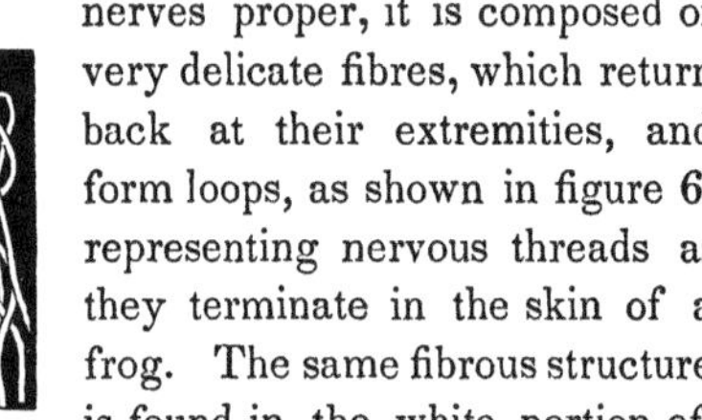
Fig. 6.

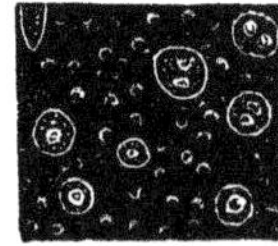
Fig. 7.

46. The tissues above enumerated differ from each other more widely, in proportion as they are examined in animals of a higher rank. As we descend in the scale of being, the differences become gradually effaced. The soft body of a snail is much more uniform in its composition than the body of a bird or a quadruped. Indeed, multitudes of animals are known to be made up of nothing but cells in contact with each other. Such is the case with the polyps; yet they contract, secrete, absorb, and reproduce; and most of the Infusoria move freely, by means of little fringes on their surface, arising from a peculiar kind of cells.

47. A no less remarkable uniformity of structure is to be observed in the higher animals, in the earlier periods of their existence, before the body has arrived at its definite form. The head of the adult salmon, for instance, contains not only all the tissues we have mentioned, namely, bone, cartilage, muscle, nerve, brain, and membranes, but also bloodvessels, glands, pigments, &c. Let us, however, examine it during the embryonic state, while it is yet in the egg, and we find that the whole head is made up of cells which differ merely in their dimensions; those at the top of the head being very small, those surrounding the eye a little larger, and those beneath being still larger, (Fig. 8.) It is only at a later period, after still further development, that these cellules become transformed, some of them into bone, others into blood, others into flesh, &c.

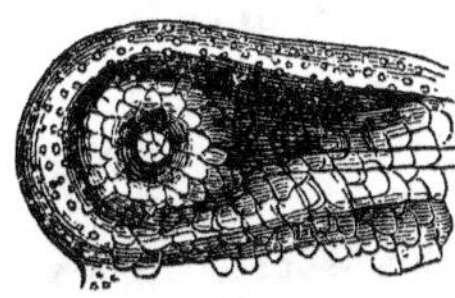

Fig. 8.

48. Again: the growth of the body, the introduction of various tissues, the change of form and structure, proceed in such a manner as to give rise to several cavities, variously combined among themselves, and each containing, at the end of these transformations, peculiar organs, or peculiar systems of organs.

SECTION III.

DIFFERENCES BETWEEN ANIMALS AND PLANTS.

49. At first glance, nothing would seem more widely different than animals and plants. What is there in common, for instance, between an oak or an elm, and the bird which seeks shelter amid their foliage ?

50. The differences are usually so obvious, that this question would be superfluous if applied only to the higher forms of the two kingdoms. But this contrast diminishes, in proportion as their structure is simplified; and as we descend to the lower forms, the distinctions are so few and so feebly characterized, that it becomes at length difficult to pronounce whether the object we have before us is an animal or a plant. Thus, the sponges have so great a resemblance to some of the polypi, that they have generally been classed among animals, although in reality they belong to the vegetable kingdom.

51. Animals and plants differ in the relative predominance of the elements, oxygen, carbon, hydrogen and nitrogen, of which they are composed. In vegetables, only a small proportion of nitrogen is found ; while it enters largely into the composition of the animal tissues.

52. Another peculiarity of the Animal Kingdom is, the presence of large, distinctly limited cavities, usually intended for the lodgment of certain organs ; such is the skull and the chest in the higher animals, the cavity of the gills in fishes, and of the abdomen, or general cavity of the body, which exists in all animals, without exception, for the purpose of digestion, or the reception of the digestive organs.

53. The well-defined and compact forms of the organs

lodged in these cavities, is a peculiarity belonging to animals only. In plants, the organs designed for special purposes are never embodied into one mass, but are distributed over various parts of the individual. Thus, the leaves, which answer to the lungs, instead of being condensed into one organ, are scattered independently in countless numbers over the branches. Nor is there any organ corresponding to the brain, the heart, the liver, or the stomach.

54. Moreover, the presence of a proper digestive cavity involves marked differences between the two kingdoms, in respect to alimentation or the use of food. In plants, the fluids absorbed by the roots are carried, through the trunk and all the branches, to the whole plant, before they arrive at the leaves, where they are to be digested. In animals, on the contrary, the food is at once received into the digestive cavity, where it is elaborated ; and it is only after it has been thus dissolved and prepared, that it is introduced into the other parts of the body. The food of animals consists of organized substances, while that of vegetables is derived from inorganic substances ; and they produce albumen, sugar, starch, &c., while animals consume them.

55. Plants commence their development from a single point, the seed, and, in like manner, all animals are developed from the egg. But the animal germ is the result of successive transformations of the yolk, while nothing similar takes place in the plant. The subsequent development of individuals is for the most part different in the two kingdoms. No limit is usually placed to the increase of plants ; trees put out new branches and new roots as long as they live. Animals, on the contrary, generally have a limited size and figure ; and these once attained, the subsequent changes are accomplished without any increase of volume, or essential alteration of form ; while the appearance of most vegetables is repeatedly modified, in a notable manner, by the develop-

ment of new branches. Some of the lowest animals, however, the polyps for instance, increase in a somewhat analogous manner, (§ 329, 330.)

56. In the effects they produce upon the air by respiration, there is an important difference. Animals consume the oxygen, and give out carbonic acid gas, which is destructive to animal life; while plants, by respiration, which they in most instances perform by means of the leaves, reverse the process, and thus furnish oxygen, which is so essential to animals. If an animal be confined in a small portion of air, or water containing air, this soon becomes so vitiated by respiration, as to be unfit to sustain life; but if living plants are enclosed with the animal at the same time, the air is maintained pure, and no difficulty is experienced. The practical effect of this compensation, in the economy of Nature, is obviously most important; vegetation restoring to the atmosphere what is consumed by animal respiration, combustion, &c., and *vice versa.*

57. But there are two things which, more than all others, distinguish the animal from the plant, namely, the power of moving itself or its parts at will, and the power of perceiving other objects or their influences; in other words, *voluntary motion* and *sensation.*

58. All animals are susceptible of undergoing pleasure and pain. Plants have also a certain sensibility. They wither and fade under a burning sun, or when deprived of moisture; and they die when subjected to too great a degree of cold, or to the action of poisons. But they have no consciousness of these influences, and suffer no pain; while animals under similar circumstances suffer. Hence they have been called *animate beings*, in opposition to plants, which are *inanimate beings.*

CHAPTER THIRD.

FUNCTIONS AND ORGANS OF ANIMAL LIFE.

SECTION I.

OF THE NERVOUS SYSTEM AND GENERAL SENSATION.

59. LIFE, in animals, is manifested by two sorts of functions, viz.: First, the peculiar *functions of animal life*, or those of *relation*, which include the functions of sensation and voluntary motion; those which enable us to approach, and perceive our fellow beings and the objects about us, and to bring us into relation with them: Second, the *functions of vegetative life*, which are nutrition in its widest sense, and reproduction;* those indeed which are essential to the maintenance and perpetuation of life.

60. The two distinguishing characteristics of animals, namely, sensation and motion, (57,) depend upon special systems of organs, which are wanting in plants, the *nervous system* and the *muscular system* under its influence. The nervous system, therefore, is the grand characteristic of the animal body. It is the centre from which all the commands of the will issue, and to which all sensations tend.

---

* This distinction is the more important, inasmuch as the organs of animal life, and those of vegetative life, spring from very distinct layers of the embryonic membrane. The first are developed from the upper layer, and the second from the lower layer of the germ of the animal. See Chapter on Embryology, p. 112.

61. Greatly as the form, the arrangement, and the volume of the nervous system vary in different animals, they may all be reduced to four principal types, which correspond, moreover, to the four great departments of the Animal Kingdom. In the vertebrate animals, namely, the fishes, reptiles, birds, and mammals, the nervous system is composed of two principal masses, the *spinal marrow*, (Fig. 9, *c*,) which runs along the back, and the *brain*, contained within the skull.* The volume of the brain is proportionally larger as the animal occupies a more elevated rank in the scale of being. Man, who stands at the head of Creation, is in this respect also the most highly endowed being.

Fig. 9.

62. With the brain and spinal marrow are connected the nerves, which are distributed, in the form of branching threads, through every part of the body. The branches which unite with the brain are twelve pairs, called the cere-

---

* The brain is composed of several distinct parts which vary greatly, in their relative proportions, in different animals, as will appear hereafter. They are — 1. The medulla oblongata; 2. Cerebellum; 3. Optic lobes; 4. Cerebral hemispheres; 5. Olfactory lobes; 6. the pituitary body; 7. the pineal body. (See figures 9 and 21.) The spinal marrow is made up by the union of four nervous columns.

bral nerves, and are designed chiefly for the organs of sense located in the head. Those which join the spinal marrow are also in pairs, one pair for each vertebra or joint of the back. The number of pairs varies, therefore, in different classes and families, according to the number of vertebræ. Each nerve is double, in fact, being composed of two threads, which at their junction with the spinal marrow are separate, and afterwards accompany each other throughout their whole course. The anterior thread transmits the commands of the will which induce motion; the other receives and conveys impressions to the brain, to produce sensations.

63. In the Articulated animals, comprising the crabs, barnacles, worms, spiders, insects, and other animals formed of rings, the nervous system consists of a series of small centres or swellings, called ganglions, (Fig. 10,) placed beneath the alimentary canal, on the floor of the general cavity of the body, and connected by threads; and of a more considerable mass placed above the œsophagus or throat, connected with the lower ganglions by threads which form a collar around the alimentary canal. The number of ganglions generally corresponds to the number of rings.

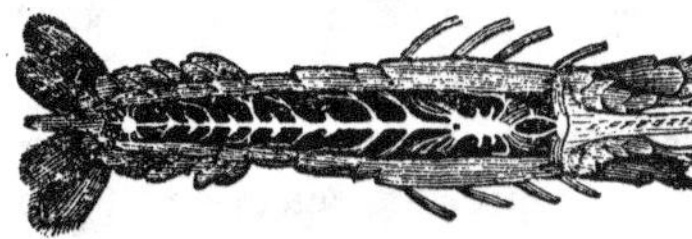
Fig. 10.

64. In the Mollusks, (Fig. 11,) the nervous system consists of a single ganglionic circle, the principal swellings of which are placed symmetrically above and below the œsophagus, and from whence the filaments, which supply the organs in different directions, take their origin.

Fig. 11.

65. In the Radiata, (Fig. 12,) the nervous system is reduced to a single ring, encircling the mouth, and giving off threads towards the circumference. It differs essentially from that of the Mollusks, by being disposed in a horizontal position, and by its star-like form.

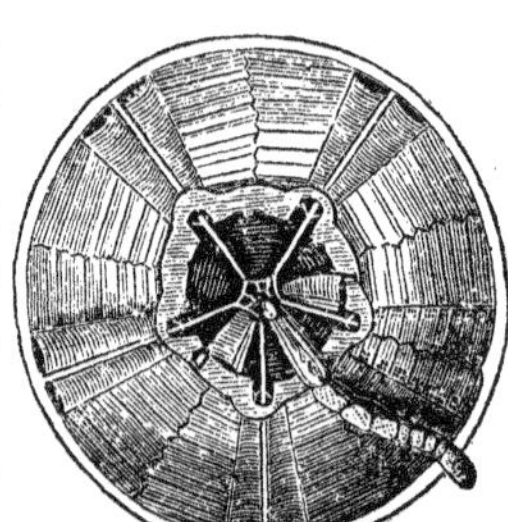
Fig. 12.

66. The nerves branch off and diffuse sensibility to every portion of the body, and thereby men and the higher animals are enabled to gain a knowledge of the general properties of the objects which surround them; every point of the body being made capable of determining whether an object is hot or cold, dry or moist, hard or soft, &c. There are some parts, however, the ends of the fingers, for example, in which this sensibility is especially acute, and these also receive a larger supply of nerves.

67. On the contrary, those parts which are destitute of sensibility, such as the feathers of birds, the wool of animals, or the hair of man, are likewise destitute of nerves. But the conclusive proof that sensibility resides in the nerves is, that when the nerve which supplies any member of the body is severed, that member at once becomes insensible.

68. There are animals in which the faculty of perception is limited to this general sense; but their number is small, and, in general, they occupy the lowest place in the series. Most animals, in addition to the general sensibility, are endowed with peculiar organs for certain kinds of perceptions, which are acted upon by certain kinds of stimuli, as light, sound and odor, and which are called the SENSES. These are five in number, namely: *sight*, *hearing*, *smell*, *taste*, and *touch*.

SECTION II.

OF THE SPECIAL SENSES.

1. *Of Sight.*

69. Sight is the sense by which light is perceived, and by means of which the outlines, dimensions, relative position, color and brilliancy of objects are discerned. Some of these properties may be also ascertained, though in a less perfect manner, by the sense of touch. We may obtain an idea of the size and shape of an object, by handling it; but the properties that have a relation to light, such as color and brilliancy, and also the form and size of bodies that are beyond our reach, can be recognized by sight only.

70. The EYE is the organ of vision. The number, structure, and position of the eyes in the body is considerably varied in the different classes. But whatever may be their position, these organs in all the higher animals are in connection with particular nerves, called the *optic nerves*, (Fig. 13, *a*.) In the vertebrates, these are the second pair of the cerebral nerves, and arise directly from the middle mass of the brain, (Fig. 21, *b*,) which, in the embryo, is the most considerable of all.

71. Throughout the whole series of vertebrate animals, the eyes are only two in number, and occupy bony cavities of the skull, called the *orbits*. The organ is a globe or hollow sphere formed by three principal membranes, enclosed one within the other, and filled with transparent matter. Figure 13 represents a vertical section

Fig. 13.

through the eye, from before backwards, and will give an dea of the relative position of these different parts.

72. The outer coat is called the *sclerotic*, (*b*;) it is a thick, firm, white membrane, having its anterior portion transparent. This transparent segment, which seems set in the opaque portion, like a watch-glass in its rim, is called the *cornea*, (*f*.)

73. The inside of the sclerotic is lined by a thin, dark-colored membrane, the *choroid*, (*c*.) It becomes detached from the sclerotic when it reaches the edge of the cornea, and forms a curtain behind it. This curtain gives to the eye its peculiar color, and is called the *iris*, (*g*.) The iris readily contracts and dilates, so as to enlarge or diminish an opening at its centre, the *pupil*, according as more or less light is desired. Sometimes the pupil is circular, as in man, the dog, the monkey; sometimes in the form of a vertical ellipse, as in the cat; or it is elongated sidewise, as in the sheep.

74. The third membrane is the *retina*, (*d*.) It is formed by the optic nerve, which enters the back part of the eye, by an opening through both the sclerotic and choroid coats, and expands upon the interior into a whitish and most delicate membrane. It is upon the retina that the images of objects are received, and produce impressions, which are conveyed by the nerve to the brain.

75. The fluids which occupy the cavity of the eye are of different densities. Behind, and directly opposite to the pupil, is placed a spheroidal body, called the *crystalline lens*, (*e*.) It is tolerably firm, perfectly transparent, and composed of layers of unequal density, the interior being always more compact than the exterior. Its form varies in different classes of animals. In general, it is more convex in aquatic than in land animals; whilst with the cornea it is directly the contrary, being flat in the former, and convex in the latter.

76. By means of the iris, the cavity, (*i*,) in front of the crys-

talline lens is divided into two compartments, called the *anterior* and *posterior chambers.* The fluid which fills these chambers is a clear watery liquid, called the *aqueous humor.* The portion of the globe behind the lens, which is much the largest, is filled by a gelatinous liquid, perfectly transparent, like that of the chambers, but somewhat more dense. This is called the *vitreous humor*, (*h.*)

77. The object of this apparatus is to receive the rays of light, which diverge from all points of bodies placed before it, and to bring them again to a point upon the retina. It is a well-known fact, that when a ray of light passes obliquely from one medium to another of different density, it will be refracted or turned out of its course more or less, according to the difference of this density, and the obliquity at which the ray strikes the surface. This may be illustrated by the following figure, (Fig. 14.)

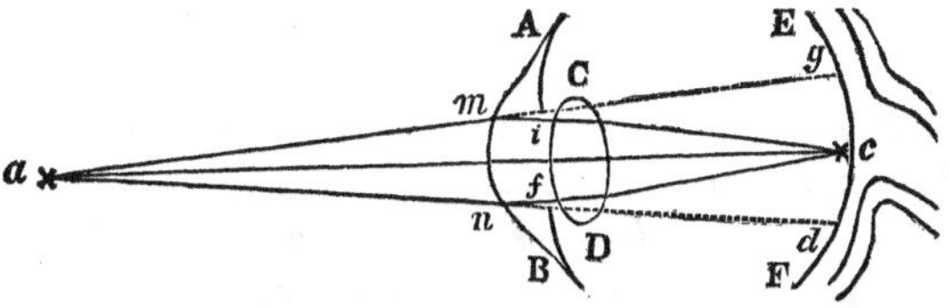

Fig. 14.

The ray *a c*, which strikes the cornea A B perpendicularly, continues without deviation, until it reaches the bottom of the eye at *c.* But the rays *a m* and *a n*, which strike the eye obliquely, change their direction, and instead of proceeding onward to *m g* and *n d*, take the direction *m i* and *n f.* A still further refraction, though less considerable, is occasioned by passing through the crystalline lens C D, and the vitreous humor, so that the two rays, *m i* and *n f*, will at last meet in a point. This point is called the *focus*, (*c*,) and in distinct vision is always precisely at the retina, E F.

78. From this arrangement, the image found upon the

retina will be inverted. We may satisfy ourselves of this by direct observation. The eye of the white rabbit being destitute of the black pigment of the choroid, is quite transparent. Take the eye, soon after the death of the animal, and arrange it in one end of a tube, so that the cornea will face outwards; then if we look in at the other end of the tube, we may see objects to which it is directed exactly pictured upon the retina, but in a reversed position.

79. The mechanical structure of the eye may be perfectly imitated by art. Indeed, the camera obscura is an instrument constructed on the very same plan. By it, external objects are pictured upon a screen, placed at the bottom of the instrument, behind a magnifying lens. The screen represents the retina; the dark walls of the instrument represent the choroid; and the cornea, the crystalline lens and the vitreous humor combined, are represented by the magnifying lens. But there is this important difference, that the eye has the power of changing its form, and of adapting itself so as to discern with equal precision very remote, as well as very near, objects.

80. By means of muscles which are attached to the ball, the eyes may be rolled in every direction, so as to view objects on all sides, without moving the head. The eyes are usually protected by lids, which are two in the mammals, and generally furnished with a range of hairs at their edges, called *eye-lashes*. Birds have a third lid, which is vertical; this is also found in most of the reptiles and a few mammals. In fishes, the lids are wanting, or immovable.

81. The eye constructed as above described is called a *simple eye*, and belongs more especially to the vertebrate animals. In man, it arrives at its highest perfection. In him, the eye also performs a more exalted office than mere vision. It is a mirror, in which the inner man is reflected. His passions, his joys, and his sorrows, his inmost self, are

revealed, with the utmost fidelity, in the expression of his eye, and it has been rightly called "the window of the soul."

82. Many of the invertebrate animals have the eye constructed upon the same plan as that of the vertebrate animals, but with this essential difference, that the optic nerve which forms the retina is not derived from a nervous centre, analogous to the brain, but arises from one of the ganglions. Thus, the eye of the cuttle-fish contains all the essential parts of the eye of the superior animals, and, what is no less important, they are only two in number, placed upon the sides of the head.

83. The snail and kindred animals have, in like manner, only two eyes, mounted on the tip of a long stalk, (the *tentacle*,) or situated at its base, or on a short pedestal by its side. Their structure is less perfect than in the cuttle-fish, but still there is a crystalline lens, and more or less distinct traces of the vitreous body. Some bivalve mollusks, the scollops for example, have likewise a crystalline lens, but instead of two eyes, they are furnished with numerous eye-spots, which are arranged like a border around the lower margin of the animal.

Fig. 15.

84. In spiders, the eyes are likewise simple, and usually eight in number. These little organs, usually called *ocelli*, instead of being placed on the sides of the body or of the head, occupy the anterior part of the back. All the essential parts of a simple eye, the cornea, the crystalline lens, the vitreous body, are found in

Fig. 16.

them, and even the choroid, which presents itself in the form of a black ring around the crystalline lens. Many insects, in their caterpillar state, also have simple eyes.

85. Rudiments of eyes have been observed in very many of the worms. They generally appear as small black spots on the head; such as are seen on the head of the Leech, the Planaria and the Nereis. In these latter animals there are four spots. According to Müller, they are small bodies, rounded behind, and flattened in front, composed of a black, cup-shaped membrane, containing a small white, opaque body, which seems to be a continuation of the optic nerve. It cannot be doubted, therefore, that these are eyes; but as they lack the optical apparatus which produces images, we must suppose that they can only receive a general impression of light, without the power of discerning objects.

86. Eye-spots, very similar to those of the Nereis, are found at the extremity of the rays of some of the star-fishes, in the sea-urchins, at the margin of many Medusæ, and in some Polypi. Ehrenberg has shown that similar spots also exist in a large number of the Infusoria.

Fig. 17.

87. In all the above-mentioned animals, the eyes, whatever their number, are apart from each other. But there is still another type of simple eyes, known as *aggregate eyes.* In some of the millipedes, the pill-bugs, for instance, the eyes are collected into groups, like those of spiders; each eye inclosing a crystalline lens and a vitreous body, surrounded by a retina and choroid. Such eyes consequently form a

natural transition to the compound eyes of insects, to which we now give our attention.

88. *Compound eyes* have the same general form as simple eyes; they are placed either on the sides of the head, as in insects, or supported on pedestals, as in the crabs. But if we examine an eye of this kind by a magnifying lens, we find its surface to be composed of an infinite number of angular, usually six-sided faces. If these façettes are removed, we find beneath a corresponding number of cones, (*c*,) side by side, five or six times as long as they are broad, and arranged like rays around the optic nerve, from which each one receives a little filament, so as to present, according to Müller, the following disposition. (Fig. 18.) The cones are perfectly transparent, but separated from each other by walls of pigment, in such a manner that only those rays which are parallel to the axes can reach the retina A; all those which enter obliquely are lost; so that of all the rays which proceed from the points *a* and *b*, only the central ones in each pencil will act upon the optic nerve, (*d*;) the others will strike against the walls of the cones. To compensate for the disadvantage of such an arrangement, and for the want of motion, the number of façettes is greatly multiplied, so that no less than 25,000 have been counted in a single eye. The image on the retina, in this case, may be compared to a mosaic, composed of a great number of small images, each of them representing a portion of the figure. The entire picture is, of course, more perfect,

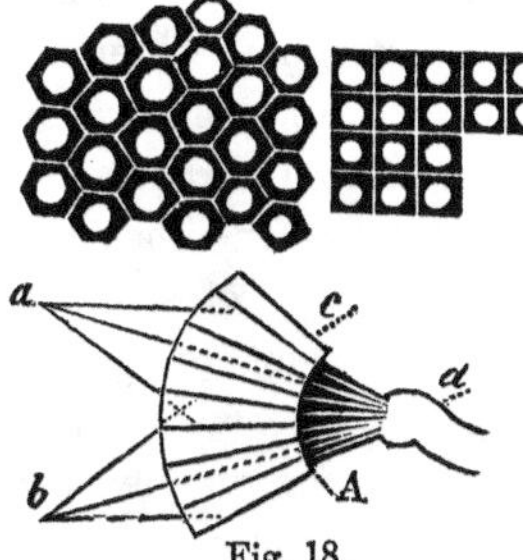

Fig. 18.

in proportion as the pieces are smaller and more numerous.

89. Compound eyes are destitute of the optical apparatus necessary to concentrate the rays of light, and cannot adapt themselves to the distance of objects; they see at a certain distance, but cannot look at pleasure. The perfection of their sight depends on the number of façettes or cones, and the manner in which they are placed. Their field of vision is wide, when the eye is prominent; it is very limited, on the contrary, when the eye is flat. Thus the dragon-flies, on account of the great prominency of their eyes, see equally well in all directions, before, behind, or laterally; whilst the water-bugs, which have the eyes nearly on a level with the head, can see to only a very short distance before them.

90. If there be animals destitute of eyes, they are either of a very inferior rank, such as most of the polypi, or else they are animals which live under unusual circumstances, such as the intestinal worms. Even among the vertebrates, there are some that lack the faculty of sight, as the *Myxine glutinosa*, which has merely a rudimentary eye concealed under the skin, and destitute of a crystalline lens. Others, which live in darkness, have not even rudimentary eyes, as, for example, that curious fish (*Amblyopsis spelæus*,) which lives in the Mammoth Cave, and which appears to want even the orbital cavity. The craw-fishes, (*Astacus pellucidus*,) of this same cave, are also blind; having merely the pedicle for the eyes, without any traces of façettes.

2. *Hearing.*

91. To hear, is to perceive sounds. The faculty of perceiving sounds is seated in a peculiar apparatus, the EAR, which is constructed with a view to collect and augment the sonorous vibrations of the atmosphere, and convey them to

the acoustic or auditory nerve, which arises from the posterior part of the brain. (Fig. 21, *c.*)

92. The ears never exceed two in number, and are placed, in all the vertebrates, at the hinder part of the head. In a large proportion of animals, as the dog, horse, rabbit, and most of the mammals, the external parts of the ear are generally quite conspicuous; and as they are, at the same time, quite movable, they become one of the prominent features of physiognomy.

93. These external appendages, however, do not constitute the organ of hearing, properly speaking. The true seat of hearing is deeper, quite in the interior of the head. It is usually a very complicated apparatus, especially in the superior animals. In mammals it is composed of three parts, the external ear, the middle ear, and the internal ear; and its structure is as follows: (Fig. 19.)

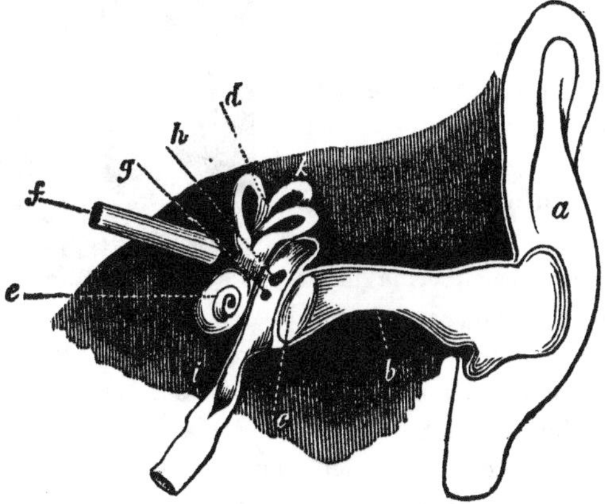

Fig. 19.

94. The external ear, which is popularly regarded as the ear, consists of the *conch*, (*a*,) and the canal which leads from it the *external auditory passage*, (*b.*) The first is a gristly expansion, in the form of a horn or a funnel, the object of which is to collect the waves of sound; for this reason, animals prick up their ears when they listen. The ear of man is remarkable for being nearly immovable. Therefore, persons, whose hearing is deficient, employ an artificial trumpet, by which the vibrations from a much more extended surface may be collected. The external ear is peculiar to mammals, and is wanting even in some aquatic species of these, such as the seals and the Ornithorhyncus.

95. The *middle ear* has received the name of the *tympanic cavity*, (*k.*) It is separated from the auditory passage by a membranous partition, the *tympanum* or drum, (*c;*) though it still communicates with the open air by means of a narrow canal, called the Eustachian tube, (*i,*) which opens at the back part of the mouth. In the interior of the chamber are four little bones, of singular forms, which anatomists have distinguished by the names of *malleus*, (Fig. 20, *c,*) *incus*, (*n,*) *stapes*, (*s,*) and *os orbiculare*, (*o;*) which are articulated together, so as to form a continuous chain, as here represented, magnified.

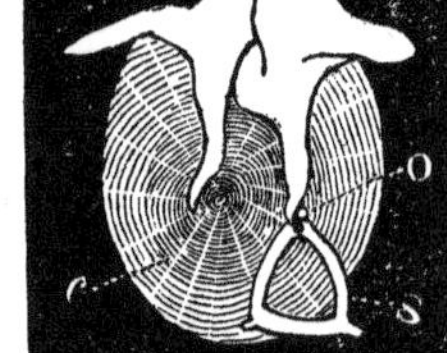

Fig. 20.

96. The *internal ear*, which is also denominated the *labyrinth*, is an irregular cavity formed in the most solid part of the temporal bone, beyond the chamber of the middle ear, from which it is separated by a bony partition, which is perforated by two small holes, called, from their form, the round and the oval apertures, the *foramen rotundum*, (Fig. 19, *g,*) and the *foramen ovale*, (*h.*) The first is closed by a membrane, similar to that of the tympanum, while the latter is closed by the *stapes*, one of the little bones in the chamber.

97. Three parts are to be distinguished in the labyrinth, namely, the *vestibule*, which is the part at the entrance of the cavity; the *semicircular canals*, (*d*,) which occupy its upper part, in the form of three arched tubes; and the *cochlea*, which is a narrow canal placed beneath, at the lower part of the vestibule, having exactly the form of a snail-shell, (*e*.) The entire labyrinth is filled with a watery fluid, in which membranous sacs or pouches float. Within these sacs, the auditory nerve (*f*) terminates. These pouches, therefore, are the actual seat of hearing, and the most essential parts of the ear. The auditory nerve is admitted to them by a long passage, the *internal auditory* canal.

98. By this mechanism, the vibrations of the air are first collected by the external ear, whence they are conveyed along the auditory passage, at the bottom of which is the tympanum. The tympanum, by its delicate elasticity, augments the vibrations, and transmits them to the internal ear, partly by means of the little bones in the chamber, which are disposed in such a manner that the stapes exactly fits the oval aperture, (*foramen ovale*;) and partly by means of the air which strikes the membrane covering the round aperture, (*g*,) and produces vibrations there, corresponding to those of the tympanum. After all these modifications, the sonorous vibrations at last arrive at the labyrinth and the auditory nerve, which transmits the impression to the brain.

99. But the mechanism of hearing is not so complicated in all classes of animals, and is found to be more and more simplified as we descend the series. In birds, the middle and interior ears are constructed on the same plans as in the mammals; but the outer ear no longer exists, and the auditory passage, opening on a level with the surface of the head behind the eyes, is merely surrounded by a circle of peculiarly formed feathers. The bones of the middle ear are also less numerous, there being generally but one.

100. In reptiles, the whole exterior ear disappears; the auditory passage is always wanting, and the tympanum becomes external. In some toads, even the middle ear also is completely wanting. The fluid of the vestibule is charged with salts of lime, which frequently give it a milky appearance, and which, when examined by the microscope, are found to be composed of an infinite number of crystals.

101. In fishes, the middle and external ear are both wanting; and the organ of hearing is reduced to a membranous vestibule, situated in the cavity of the skull, and surmounted by semicircular canals, from one to three in number. The liquid of the vestibule contains chalky concretions of irregular forms, which are called Otolites, the use of which is doubtless to render the vibration of sounds more sensible.

102. In crabs, the organ of hearing is found on the lower face of the head, at the base of the large antennæ. It is a bony chamber closed by a membrane, in the interior of which is suspended a membranous sac filled with water. On this sac, the auditory nerve is expanded. In the cuttlefish, the vestibule is a simple excavation of the cartilage of the head, containing a little membranous sac, in which the auditory nerve terminates.

103. Finally, some insects, the grasshopper for instance, have an auditory apparatus, no longer situated in the head, as with other animals, but in the legs; and from this fact, we may be allowed to suppose, that if no organ of hearing has yet been found in most insects, it is because it has been sought for in the head only.

104. It appears from these examples, that the part of the organ of hearing which is uniformly present in all animals furnished with ears, is precisely that in which the auditory nerve ends. This, therefore, is the essential part of the organ. The other parts of the apparatus, the tympanum, auditory passage, and even the semicircular canals, have for

their object merely to aid the perception of sound with more precision and accuracy. Hence we may conclude that the sense of hearing is dull in animals where the organ is reduced to its most simple form; and that animals which have merely a simple membranous sac, without tympanum and auditory passage, as the fishes, or without semicircular canals, as the crabs, perceive sounds in but a very imperfect manner.

3. *Of Smell.*

105. SMELL is the faculty of perceiving odors, and is a highly important sense to many animals. Like sight and hearing, smell depends upon special nerves, the *olfactory*, (*a*,) which are the first pair of cerebral nerves, and which, in the embryo, are direct prolongations of the brain.

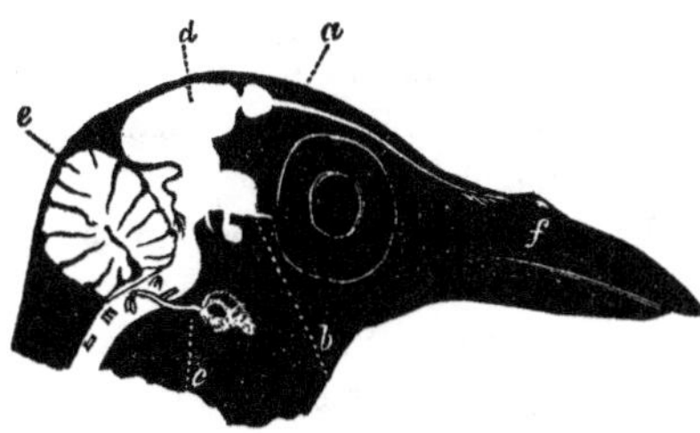

Fig. 21.

a, olfactory nerve; *b*, optic nerve; *c*, auditory nerve; *d*, cerebrum; *e*, cerebellum; *f*, nostril.

106. The organ of smell is the NOSE. Throughout the series of vertebrates, it makes a part of the face, and in man, by reason of its prominent form, it becomes one of the dominant traits of his countenance; in other mammals, the nose loses this prominency by degrees, and the nostrils no longer open downwards, but forwards. In birds, the position of the nostrils is a little different; they open farther back and higher, at the origin of the beak,.(*f*.)

107. The nostrils are usually two in number. Some fishes have four. They are similar openings, separated by a partition upon the middle line of the body. In man and the

mammals, the outer walls of the nose are composed of cartilage; but internally, the nostrils communicate with bony cavities situated in the bones of the face and forehead. These cavities are lined by a thick membrane, the *pituitary* membrane, on which are expanded the nerves of smell, namely, the olfactory nerves, and some filaments of the nerve which goes to the face.

108. The process of smelling is as follows. Odors are particles of extreme delicacy which escape from very many bodies, and are diffused through the air. These particles excite the nerves of smell, which transmit the impressions made on them to the brain. To facilitate the perception of odors, the nostrils are placed in the course of the respiratory passages, so that all the odors which are diffused in the air inspired, pass over the pituitary membrane.

109. The acuteness of the sense of smell depends on the extent to which the membrane is developed. Man is not so well endowed in this respect as many animals, which have the internal surface of the nostrils extremely complicated, as it is especially among the beasts of prey.

110. The sense of smell in Reptiles is less delicate than in the mammals; the pituitary membrane, also, is less developed. Fishes are probably still less favored in this respect. As they perceive odors through the medium of water, we should anticipate that the structure of their apparatus would be different from that of animals which breathe in the air. Their nostrils are mere superficial pouches, lined with a membrane gathered into folds which generally radiate from a centre, but are sometimes arranged in parallel ridges on each side of a central band. As the perfection of smell depends on the amount of surface exposed, it follows that those fishes which have these folds most multiplied are also those in which this sense is most acute.

111. No special apparatus for smell has yet been found in Invertebrates. And yet there can be no doubt that insects, crabs, and some mollusks perceive odors, since they are attracted from a long distance by the odor of objects. Some of these animals may be deceived by odors similar to those of their prey; which clearly shows that they are led to it by this sense. The carrion fly will deposit its eggs on plants which have the smell of tainted flesh.

4. *Of Taste.*

112. TASTE is the sense by which the flavor of bodies is perceived. That the flavor of a body may be perceived, it must come into immediate contact with the nerves of taste; these nerves are distributed at the entrance to the digestive tube, on the surface of the tongue and the palate. By this sense, animals are guided in the choice of their food, and warned to abstain from what is noxious. There is an intimate connection between the taste and the smell, so that both these senses are called into requisition in the selection of food.

113. The nerves of taste are not so strictly special as those of sight and hearing. They do not proceed from one single trunk, and, in the embryo, do not correspond to an isolated part of the brain. The tongue, in particular, receives nerves from several trunks; and taste is perfect in proportion as the nerves which go to the tongue are more minutely distributed. The extremities of the nerves generally terminate in little asperities of the surface, called *papillæ*. Sometimes these papillæ are very harsh, as in the cat and the ox; and again they are very delicate, as in the human tongue, in that of the dog, horse, &c.

114. Birds have the tongue cartilaginous, sometimes beset with little stiff points; sometimes fibrous or fringed at the edges. In the parrots, it is thick and fleshy;

or it is even barbed at its point, as in the woodpeckers. In some reptiles, the crocodile for example, the tongue is adherent; in others, on the contrary, it is capable of extensive motion, and serves as an organ of touch, as in the serpents, or it may be thrust out to a great length to take prey, like that of the chameleon, toad, and frog. In fishes, it is usually cartilaginous, as in birds, generally adherent, and its surface is frequently covered with teeth.

115. It is to be presumed, that in animals which have a cartilaginous tongue, the taste must be very obtuse, especially in those which, like most fishes, and many granivorous birds, swallow their prey without mastication. In fishes, especially, the taste is very imperfect, as is proved by their readily swallowing artificial bait. It is probable that they are guided in the choice of their prey by sight, rather than by taste or smell.

116. Some of the inferior animals select their food with no little discernment. Thus, flies will select the sugary portions of bodies. Some of the mollusks, as the snails for example, are particularly dainty in the choice of their food. In general, the taste is but imperfectly developed, except in the mammals, and they are the only animals which enjoy the flavor of their food. With man, this sense, like others, may be greatly improved by exercise; and it is even capable of being brought to a high degree of delicacy.

5. *Of Touch.*

117. The sense of TOUCH is merely a peculiar manifestation of the general sensibility, seated in the skin, and dependent upon the nerves of sensation, which expand over the surface of the body. By the aid of this general sensibility, we learn whether a body is hot or cold, wet or dry. We may also, by simple contact, gain an idea, to a certain extent, of the form and consistence of a body, as, for example, whether it be sharp or blunt, soft or hard.

118. This faculty resides more especially in the hand, which is not only endowed with a more delicate tact, but, owing to the disposition of the fingers, and the opposition of the thumb to the other fingers, is capable of so moulding itself around objects, as to multiply the points of contact. Hence, touch is an attribute of man, rather than of other animals; for among these latter, scarcely any, except the monkeys, have the faculty of touch in their hands, or, as it is technically termed, of *palpation.*

119. In some animals, this faculty is exercised by other organs. Thus the trunk of the elephant is a most perfect organ of touch; and probably the mastodon, whose numerous relics are found scattered in the superficial layers of the earth's crust, was furnished with a similar organ. Serpents make use of their tongue for touch; insects employ their palpi, and snails their tentacles, for the same purpose.

6. *The Voice.*

120. Animals have not only the power of perceiving, but many of them have also the faculty of producing sounds of every variety, from the roaring of the lion to the song of the bird as it salutes the rising sun. It is moreover to be remarked that those which are endowed with a voice, likewise have the organ of hearing well developed.

121. Animals employ their voice either for communication with each other, or to express their sensations, their enjoyments, their sufferings. Nevertheless, this faculty is enjoyed by but a small minority of animals; with but very few exceptions, only the mammals, the birds, and a few reptiles are endowed with it. All others are dumb. Worms and insects have no true voice; for we must not

mistake for it the buzzing of the bee, which is merely a noise created by the vibration of the wings; nor the grating shriek of the Locust, (grasshopper,) caused by the friction of his legs against his wings; nor the shrill noises of the cricket, or the tell-tale call of the katydid, produced by the friction of the wing covers upon each other, and in numerous similar cases which might be cited.

122. Consequently, were the mammals, the birds, and the frogs to be struck out of existence, the whole Animal Kingdom would be dumb. It is difficult for us, living in the midst of the thousand various sounds which strike our ear from all sides, to conceive of such a state. Yet such a state did doubtless prevail for thousands of ages, on the surface of our globe, when the watery world alone was inhabited, and before man, the birds, and the mammals were called into being.

123. In man and the mammals, the voice is formed in an organ called the *larynx*, situated at the upper part of the windpipe, below the bone of the tongue, (*a*.) The human larynx, the part called Adam's apple, is composed of several cartilaginous pieces, called the thyroid cartilage, (*b*,) the cricoid cartilage, (*c*,) and the small arytenoid cartilages. Within these are found two large folds of elastic substance, known by the name of the *vocal cords*, (*m*.) Two other analogous folds, the *superior ligaments of the glottis*, (*n*,) are situated a little above the preceding. The glottis (*o*) is the space between these four folds. The arrangement of the vocal cords, and of the interior of the glottis in man, is indicated by dotted lines, in Fig. 22.

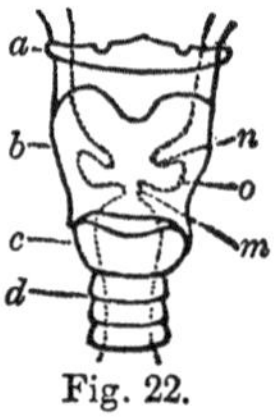

Fig. 22.

124. The mechanism of the voice is as follows: the air, on its way to the lungs, passes the vocal cords. So long as these are in repose, no sound is produced; but the moment they are made tense they narrow the aperture, and oppose

an obstacle to the current of air, and it cannot pass without causing them to vibrate. These vibrations produce the voice; and as the vocal cords are susceptible of different degrees of tension, these tensions determine different sounds; giving an acute tone when the tension is great, but a grave and dull one when the tension is feeble.

125. Some mammals have, in addition, large cavities which communicate with the glottis, and into which the air reverberates, as it passes the larynx. This arrangement is especially remarkable in the howling monkeys, which are distinguished above all other animals for their deafening howls.

126. In birds, the proper larynx is very simple, destitute of vocal cords, and incapable of producing sounds; but at the lower end of the windpipe there is a second or inferior larynx, which is very complicated in structure. It is a kind of bony drum, (*a*,) having within it two glottides, formed at the top of the two branches (*bb*) of the windpipe, (*c*,) each provided with two vocal cords. The different pieces of this apparatus are moved by peculiar muscles, the number of which varies in different families. In birds which have a very monotonous cry, such as the gulls, the herons, the cuckoos, and the mergansers, (Fig. 23,) there is but one or two pairs; parrots have three; and the birds of song have five.

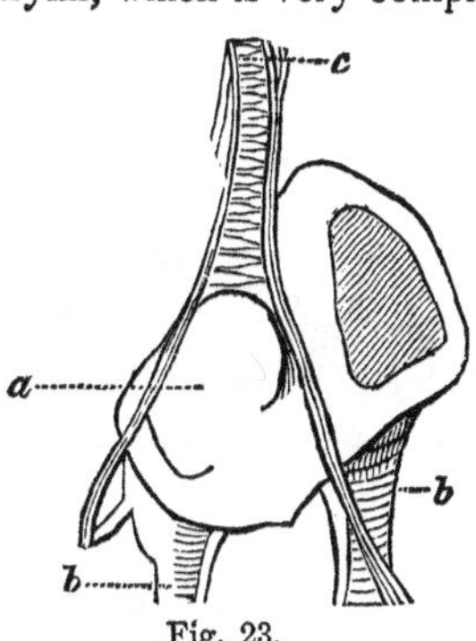

Fig. 23.

127. Man alone, of all the animal creation, has the power of giving to the tones he utters a variety of definite or articulate sounds; in other words, he alone has the gift of speech.

CHAPTER FOURTH.

OF INTELLIGENCE AND INSTINCT.

128. Besides the material substance of which the body is constructed, there is also an immaterial principle, which, though it eludes detection, is none the less real, and to which we are constantly obliged to recur in considering the phenomena of life. It originates with the body, and is developed with it, while yet it is totally apart from it. The study of this inscrutable principle belongs to one of the highest branches of Philosophy; and we shall here merely allude to some of its phenomena which elucidate the development and rank of animals.

129. The *constancy of species* is a phenomenon depending on the immaterial nature. Animals, and plants also, produce their kind, generation after generation. We shall hereafter show that all animals may be traced back, in the embryo, to a mere point in the yolk of the egg, bearing no resemblance whatever to the future animal; and no inspection would enable us to declare with certainty what that animal is to be. But even here an immaterial principle is present, which no external influence can essentially modify, and determines the growth of the future being. The egg of the hen, for instance, cannot be made to produce any other animal than a chicken, and the egg of the codfish produces only the cod. It may therefore be said with truth, that the chicken and the cod existed in the egg before their formation as such.

130. Perception is a faculty springing from this principle. The organs of sense are the instruments for receiving

sensations, but they are not the faculty itself, without which they would be useless. We all know that the eye and ear may be open to the sights and sounds about us; but if the mind happens to be preoccupied, we perceive them not. We may even be searching for something which actually lies within the compass of our vision; the light enters the eye as usual, and the image is formed on the retina; but, to use a common expression, we look without seeing, unless the mind that perceives is directed to the object.

131. In addition to the faculty of perceiving sensations, the higher animals have also the faculty of recalling past impressions, or the power of *memory*. Many animals retain a recollection of the pleasure or pain they have experienced, and seek or avoid the objects which may have produced these sensations; and, in doing so, they give proof of *judgment*.

132. This fact proves that animals have the faculty of comparing their sensations and of deriving conclusions from them; in other words, that they carry on a process of *reasoning*.

133. These different faculties, taken together, constitute *intelligence*. In man, this superior principle, which is an emanation of the divine nature, manifests itself in all its splendor. God "breathed into him the breath of life, and man became a living soul." It is man's prerogative, and his alone, to regulate his conduct by the deductions of reason; he has the faculty of exercising his judgment not only upon the objects which surround him, and of apprehending the many relations which exist between himself and the external world; he may also apply his reason to immaterial things, observe the operations of his own intellect, and, by the analysis of his faculties, may arrive at the consciousness of his own nature, and even conceive of that Infinite Spirit, "whom none by searching can find out."

134. Other animals cannot aspire to conceptions of this kind, they perceive only such objects as immediately strike their senses, and are incapable of continuous efforts of the reasoning faculty in regard to them. But their conduct is frequently regulated by another principle of inferior order, still derived from the immaterial principle, called INSTINCT.

135. Under the guidance of Instinct, animals are enabled to perform certain operations, without instruction, in one undeviating manner. When man chooses wood and stone, as the materials for his dwelling, in preference to straw and leaves, it is because he has learned by experience, or because his associates have informed him, that these materials are more suitable for the purpose. But the bee requires no instructions in building her comb. She selects at once the fittest materials, and employs them with the greatest economy; and the young bee exhibits, in this respect, as much discernment as those who have had the benefit of long experience. She performs her task without previous study, and, to all appearances, without the consciousness of its utility, being in some sense impelled to it by a blind impulse.

136. If, however, we judge of the instinctive acts of animals when compared with acts of intelligence, by the relative perfection of their products, we may be led into gross errors, as a single example will show. No one will deny that the honey-comb is constructed with more art and care than the huts of many tribes of men. And yet, who would presume to conclude from this that the bee is superior in intelligence to the inhabitant of the desert or of the primitive forest? It is evident, on the contrary, that in this particular case we are not to judge of the artisan by his work. As a work of man, a structure as perfect in all respects as the honey-comb would indicate very complicated mental operations, and probably would require numerous preliminary experiments.

137. The instinctive actions of animals relate either to

the procuring of food, or to the rearing of their young; in other words, they have for their end the preservation of the individual and of the species. It is by instinct that the leopard conceals himself and awaits the approach of his prey. It is equally by instinct that the spider spreads his web to entangle the flies which approach it.

138. Some animals go beyond these immediate precautions; their instinct leads them to make provision for the future. Thus the squirrel lays in his store of nuts and acorns during autumn, and deposits them in cavities of trees, which he readily finds again in winter. The hamster digs, by the side of his burrow, compartments for magazines, which he arranges with much art. Finally, the bee, more than any other animal, labors in view of the future; and she has become the emblem of order and domestic economy.

139. Instinct exhibits itself, in a no less striking manner, in the anxiety which animals manifest for the welfare of their anticipated progeny. All birds build nests for the shelter and nurture of their young, and in some cases these nests are made exceedingly comfortable. Others show very great ingenuity in concealing their nests from the eyes of their enemies, or in placing them beyond their reach. There is a small bird in the East Indies, the tailor bird, (Sylvia sutoria,) which works wool or cotton into threads, with its feet and beak, and uses it to sew together the leaves of trees for its nest.

140. The nest of the fiery hang-bird, (Icterus Baltimore,) dangling from the extremity of some slender, inaccessible twig, is familiar to all. The beautiful nest of the humming-bird, seated on a mossy bough, and itself coated with lichen and lined with the softest down from the cotton-grass or the mullein leaf, is calculated equally for comfort and for escaping observation. An East Indian bird, (Ploceus Philippinus,) not only exhibits wonderful devices in the construction,

security, and comfort of its nest, but displays a still further advance towards intelligence. The nest is built at the tips of long pendulous twigs, usually hanging over the water. It is composed of grass, in such a manner as to form a complete thatch. The entrance is through a long tube, running downwards from the edge of the nest; and its lower end is so loosely woven, that any serpent or squirrel, attempting to enter the aperture, would detach the fibres, and fall to the ground. The male, however, who has no occasion for such protection, builds his thatched dome, similar to that of the female, and by its side; but makes simply a perch across the base of the dome, without the nest-pouch or tube.

Fig. 24.

141. But it is among insects that this instinctive solicitude for the welfare of the progeny is every where exhibited in the most striking manner. Bees and wasps not only prepare cells for each of their eggs, but take care, before closing the cells, to deposit in each of them something appropriate for the nourishment of the future young.

142. It is by the dictate of instinct, also, that vast numbers of animals of the same species associate, at certain periods of the year, for migration from one region to another; as the swallows and passenger pigeons, which are sometimes met with in countless flocks.

143. Other animals live naturally in large societies, and labor in common. This is the case with the ants and bees. Among the latter, even the kind of labor for each member of the community is determined beforehand, by instinct.

Some of them collect only honey and wax; while others are charged with the care and education of the young; and still others are the natural chiefs of the colony..

144. Finally, there are certain animals so guided by their instinct as to live like pirates, on the avails of others' labor. The Lestris or Jager will not take the trouble to catch fish for itself, but pursues the gulls, until, worn out by the pursuit, they eject their prey from their crop. Some ants make war upon others less powerful, take their young away to their nests, and oblige them to labor in slavery.

145. There is a striking relation between the volume of the brain compared with the body, and the degree of intelligence which an animal may attain. The brain of man is the most voluminous of all, and among other animals there is every gradation in this respect. In general, an animal is the more intelligent, in proportion as its brain bears a greater resemblance to that of man.

146. The relation between instinct and the nervous system does not present so intimate a correspondence as exists between the intellect and the brain. Animals which have a most striking development of instinct, as the ants and bees, belong to a division of the Animal Kingdom where the nervous system is much less developed than that of the vertebrates, since they have only ganglions, without a proper brain. There is even a certain antagonism between instinct and intelligence, so that instinct loses its force and peculiar character, whenever intelligence becomes developed.

147. Instinct plays but a secondary part in man. He is not, however, entirely devoid of it. Some of his actions are entirely prompted by instinct, as, for instance, the attempts of the infant to nurse. The fact, again, that these instinctive actions mostly belong to infancy, when intelligence is but slightly developed, goes to confirm the two last propositions.

CHAPTER FIFTH.

OF MOTION.

SECTION I.

APPARATUS OF MOTION.

148. THE power of voluntary motion is the second grand characteristic of animals, (57.) Though they may not all have the means of transporting themselves from place to place, there is no one which has not the power of executing some motions. The oyster, although fixed to the ground, opens and closes its shell at pleasure; and the little coral animal protrudes itself from its cell, and retires again at its will.

149. The movements of animals are effected by means of *muscles*, which are organs designed expressly for this purpose, and which make up that portion of the body which is commonly called *flesh*. They are composed of threads, which are readily seen in boiled meat. These threads are again composed of still more delicate fibres, called *muscular fibres*, (45,) which have the property of elongating and contracting.

150. The motions of animals and plants depend, therefore, upon causes essentially different. The expansion and closing of the leaves and blossoms of plants, which are their most

obvious motions, are due to the influence of ligh , heat, moisture, cold, and similar external agents ; but all the motions peculiar to animals are produced by a cause residing within themselves, namely, the *contractility* of muscular fibres.

151. The cause which excites contractility resides in the nerves, although its nature is not precisely understood. We only know that each muscular bundle receives one or more nerves, whose filaments pass at intervals across the muscular fibres, as seen in Fig. 25. It has also been shown, by experiment, that when a nerve entering a muscle is severed, the muscle instantly loses its power of contracting under the stimulus of the will, or, in other words, is paralyzed.

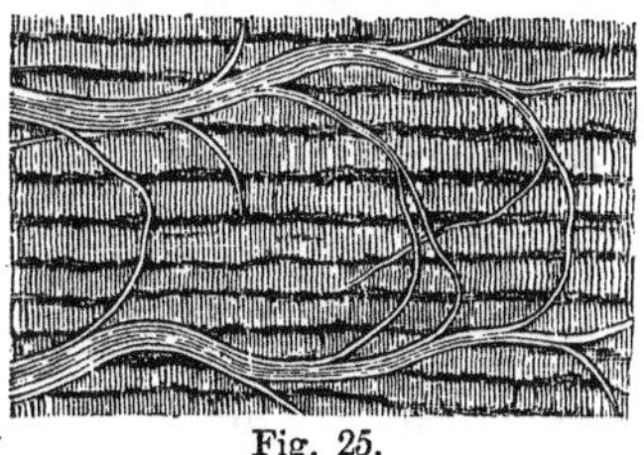
Fig. 25.

152. The muscles may be classified, according as they are more or less under the control of the will. The contractions of some of them are entirely dependent on the will, as in the muscles of the limbs used for locomotior. Others are quite independent of it, like the contractions of the heart and stomach. The muscles of respiration ordinarily act independently of the will, but are partially subject to it: thus, when we attempt to hold the breath, we arrest, for the moment, the action of the diaphragm.

153. In the great majority of animals, motion is greatly aided by the presence of solid parts, of a bony or horny structure, which either serve as firm attachments to the muscles, or, being arranged so as to act as levers, to increase the precision and sometimes the force of movements. The solid parts are usually so arranged as to form a sub-

stantial framework for the body, which has been variously designated in the several classes of animals, as the *test*, *shell carapace*, *skeleton*, *&c.* The study of these parts is one of the most important branches of comparative anatomy. Their characters are the most constant and enduring of all others. Indeed, these solid parts are nearly all that remains of the numerous extinct races of animals of past geological eras; and from these alone are we to determine the structure and character of the ancient fauna.

154. Most of the Radiata have a calcareous test or crusty shell. In the Polypi, this structure, when it exists, is usually very solid, sometimes assuming the form of a simple internal skeleton, or forming extensively branched stems, as in the sea-fans; or giving rise to solid masses, furnished with numerous cavities opening at the surface, from which the movable parts of the animals are protruded, with the power, however, of retracting themselves at pleasure, as in the corals. In the Echinoderms, the test is intimately connected with the structure of the soft parts. It is composed of numerous little plates, sometimes consolidated and immovable, as in the sea-urchins, (Fig. 26,) and sometimes so combined, as to allow of various motions, as in the star-fishes, (Fig. 17,) which use their projecting rays, both for crawling and swimming.

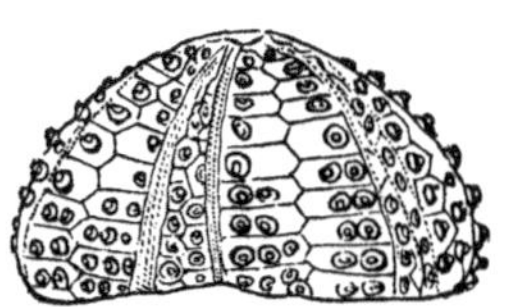

Fig. 26.

155. In the Mollusks, the solid parts are secreted by the skin, most frequently in the form of a calcareous shell of one, two, or many pieces, serving for the protection of the soft parts which they cover. These shells are generally so constructed as to afford complete protection to the animal within their cavities. In a few, the shell is too small for this purpose; and in some it exists only at a very early period,

and is lost as the animal is developed, so that at last .here is no other covering than a slimy skin. In others, the skin becomes so thick and firm as to have the consistence of elastic leather; or it is gelatinous or transparent, and, what is very curious, these tissues may be the same as those of woody fibre, as, for example, in the Ascidia. As a general thing, the solid parts do not aid in locomotion, so that the mollusks are mostly sluggish animals. It is only in a few rare cases that the shell becomes a true lever, as in the Scollops, (Pecten,) which use their shells to propel themselves in swimming.

156. The muscles of mollusks either form a flat disk under the body, or large bundles across its mass, or are distributed in the skin so as to dilate and contract it, or are arranged about the mouth and tentacles, which they put in motion. However varied the disposition may be, they always form very considerable masses, in proportion to the size of the body, and have a soft and mucous appearance, such as is not seen in the contractile fibres of other animals. This peculiar aspect no doubt arises from the numerous small cavities extending between the muscles, and the secretion of mucus which takes place in them.

157. In the Articulated animals, the solid parts are external, in the form of rings, generally of a horny structure, but sometimes calcareous, and successively fitting into each other at their edges. The tail of a lobster gives a good idea of this structure. The rings differ in the severa classes of this department, merely as to volume, form, solidity, number of pieces, and the degree of motion which one has upon another. In some groups they are consolidated, so as to form a shield or carapace, such as we see in the crabs. In others, they are membranous, and the body is capable of assuming various forms, as in the leeches and worms generally

158. A variety of appendages are attached to these rings, such as jointed legs, or in place of them stiff bristles, oars fringed with silken threads, wings either firm or membranous, antennæ, movable pieces which perform the office of jaws, &c. But however diversified this solid apparatus may be, it is universally the case that the rings, to which every segment of the body may be referred as to a type, combine to form but a single internal cavity, in which all the organs are enclosed, the nervous system, as well as the organs of vegetative life, (63.)

159. The muscles which move all these parts have this peculiarity, that they are all enclosed within the more solid framework, and not external to it, as in the vertebrates; and also that the muscular bundles, which are very considerable in number, have the form of ribbons, or fleshy strips, with parallel fibres of remarkable whiteness. Figure 27 represents the disposition of the muscles of the caterpillar which destroys the willow, (*Cossus ligniperda.*) The right side represents the superficial layer of muscles, and the left side the deep-seated layer.

Fig. 27.

160. The Vertebrata, like the articulated animals, have solid parts at the surface, as the hairs and horns of mammals, the coat of mail of the armadillo, the feathers and claws of birds, the bucklers and scales of reptiles and fishes, &c. But they have besides this, along the interior of the whole body, a solid framework not found in the invertebrates, well known as the SKELETON.

161. The skeleton is composed of a series of separate bones, called vertebræ, united to each other by ligaments.

Each vertebra has a solid centre with four branches, two of which ascend and form an arch above, and two descend, forming an arch below the body of the vertebra. The upper arches form a continuous cavity (*a*) along the region of the trunk, which encloses the spinal marrow, and in the head receives the brain, (61.) The lower arches (*b*) form another cavity, similar to the superior one, which contains the organs of nutrition and reproduction; their branches generally meet below, and when disjoined, the deficiency is supplied by fleshy walls. Every part of the skeleton may be reduced to this fundamental type the vertebra, as will be shown, when treating specially of the vertebrate animals; so that between the pieces composing the head, the trunk, or the tail, we have only differences in the degree of development of the body of the vertebra, or of its branches, and not in reality different plans of organization.

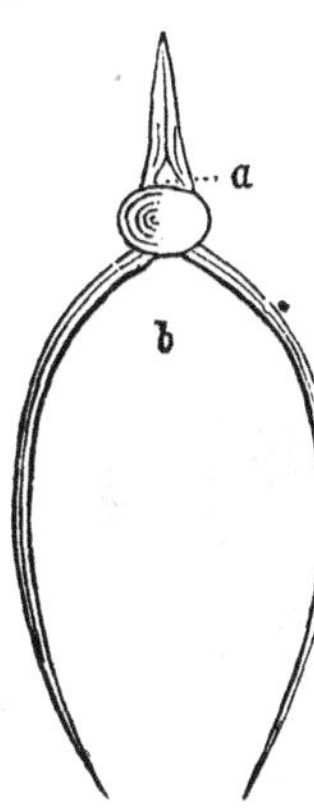

Fig. 28.

162. The muscles which move this solid framework of the vertebrata are disposed around the vertebræ, as is well exemplified among the fishes, where there is a band of muscles for each vertebra. In proportion as limbs

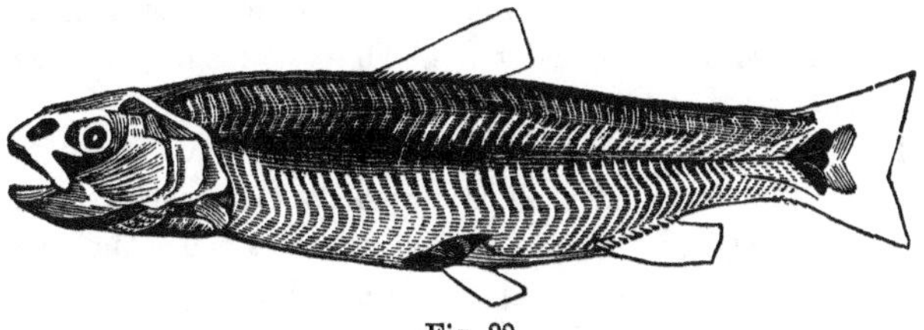
Fig. 29.

are developed, this intimate relation between the muscles and the vertebræ diminishes. The muscles are unequally distributed and are concentrated about the limbs, where the greatest amount of muscular force is required. For this reason, the largest masses of flesh in the higher vertebrates are found about the shoulders and hips; while in fishes they are concentrated about the base of the tail, which is the part principally employed in locomotion.

Fig. 30.

SECTION II.

OF LOCOMOTION.

163. One of the most curious and important applications of this apparatus of bones and muscles is for Locomotion. By this is understood the movement which an animal makes in passing from place to place, in the pursuit of pleasure, sustenance, or safety, in distinction from those motions which are performed equally well while stationary, such as the acts of respiration, mastication, &c.

164. The means which nature has brought into action to effect locomotion under all the various circumstances in which animals are placed, are very diversified; and the study of their adaptation to the necessities of animals is highly interesting in a mechanical, as well as in a zoölogical point of view. Two general plans may be noticed, under which these varieties may be arranged. Either the whole body is

equally concerned in effecting locomotion, or only some of its parts are employed for the purpose.

Fig. 31.

165. The jelly-fishes (Medusæ) swim by contracting their umbrella-shaped bodies upon the water below, and its resistance urges them forwards. Other animals are provided with a sac or siphon, which they may fill with water, and suddenly force out, producing a jet, which is resisted by the surrounding water, and the animal is thus propelled. The Bîche-le-mar, (Holothuria,) the cuttle-fishes, the Salpæ, &c., move in this way.

166. Others contract small portions of the body in succession, which being thereby rendered firmer, serve as points of resistance, against which the animal may strive, in urging the body onwards. The earth-worm, whose body is composed of a series of rings united by muscles, and shutting more or less into each other, has only to close up the rings at one or more points, to form a sort of fulcrum, against which the rest of the body exerts itself in extending forwards.

167. Some have, at the extremities of the body, a cup or some other organ for maintaining a firm hold, each extremity acting in turn as a fixed point. Thus the Leech has a cup or sucker at its tail, by which it fixes itself; the body is then elongated by the contraction of the muscular fibres which encircle the animal; the mouth is next fixed by a similar sucker and by the contraction of muscles running lengthwise the body is shortened, and the tail, losing its hold, is brought forwards to repeat the same process. Most of the bivalve mollusks, such as the clams,

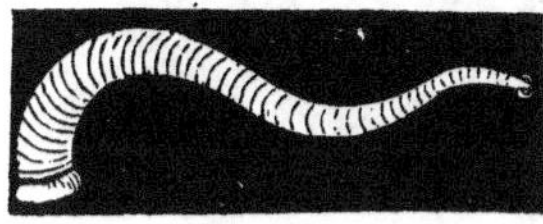

Fig. 32.

move from place to place, in a similar way. A fleshy organ, called the foot, is thrust forward, and its extremity fixed in the mud, or to some firm object, when it contracts, and thus draws along the body and the shell enclosing it. Snails, and many similar animals, have the fleshy under surface of their body composed of an infinitude of very short muscles, which, by successive contractions, so minute, indeed, as scarcely to be detected, enable them to glide along smoothly and silently, without any apparent muscular effort.

168. In the majority of animals, however, locomotion is effected by means of organs specially designed for the purpose. The most simple are the minute, hair-like *cilia*, which fringe the body of most of the microscopic infusory animalcules, and which, by their incessant vibrations, cause rapid movements. The sea-urchins and star-fishes have little thread-like tubes issuing from every side of the body, furnished with a sucker at the end. By attaching these to some fixed object, they are enabled to draw or roll themselves along; but their progress is always slow. Insects are distinguished for the number and great perfection of their organs of motion. They have at least three pairs of legs, and usually wings also. But those that have numerous feet, like the centipedes, are not distinguished for agility. The Crustacea generally have at least five pairs of legs, which are used for both swimming and crawling. The Worms are much less active; some of them have only short bristles at their sides. Some of the marine species use their fringe-like gills for paddles. (Fig. 33.)

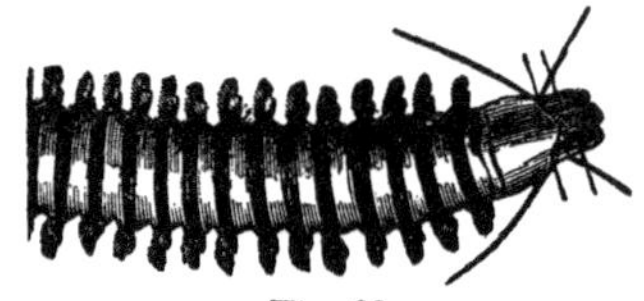
Fig. 33.

169. Among the Vertebrata, we find the greatest diversity in the organs of locomotion and the modes of their application, as well as the greatest perfection, in whatever element

they may be employed. The sailing of the eagle, the bounding of the antelope, the swimming of the shark, are not equalled by any movements of insects. This superiority is due to the internal skeleton, which, while it admits a great display of force, gives to the motions, at the same time, a great degree of precision.

1. *Plan of the Organs of Locomotion.*

170. The organs of progression in vertebrated animals never exeeed four in number, and to them the term *limbs* is more particularly applied. The study of these organs, as characteristic of the different groups of vertebrate animals, is most interesting, especially when prosecuted with a view to trace them all back to one fundamental plan, and to observe the modifications, oftentimes very slight, by which a very simple organ is adapted to every variety of movement. No part of the animal structure more fully illustrates the unity of design, or the skill of the Intellect which has so adapted a single organ to such multiplied ends. On this account, we shall illustrate this subject somewhat in detail.

171. It is easy to see that the wing which is to sustain the bird in the air must be different from the leg of the stag, which is to serve for running, or the fins of the fish that swims. But, notwithstanding their dissimilarity, the wing of the bird, the leg of the stag, and the shoulder fin of the fish, may still be traced to the same plan of structure; and if we examine their skeletons, we find the same fundamental parts. In order to show this, it is necessary to give a short description of the composition of the arm or anterior extremity.

172. The anterior member, in the vertebrates, is invariably composed of the following bones: 1. The *shoulder-blade*, or *scapula*, (*a*,) a broad and flat bone, applied upon the bones of the trunk. 2. The *arm*, (*b*,) formed of a single

long cylindrical bone, the *humerus;* 3. The *fore-arm* composed of two long bones, the *radius*, (*c*,) and *ulna*, (*d*,) which are often fused into one; 4. The *hand*, which is composed of a series of bones, more or less numerous in different classes, and which is divided into three parts, namely, the *carpus*, or wrist, (*e*,) the *metacarpus*, or palm, (*f*,) and the *phalanges*, or fingers, (*g*.) The clavicle or *collar-bone*, (*o*,) when it exists, belongs also to the anterior member. It is a bone of a cylindrical form, fixed as a brace between the breast-bone and shoulder-blade. Its use is to keep the shoulders separated; to this end, we find it fully developed in all animals which raise the limbs from the sides, as the birds and the bats. On the other hand, it is rudimentary, or entirely wanting in animals which move them backwards and forwards only, as with most quadrupeds.

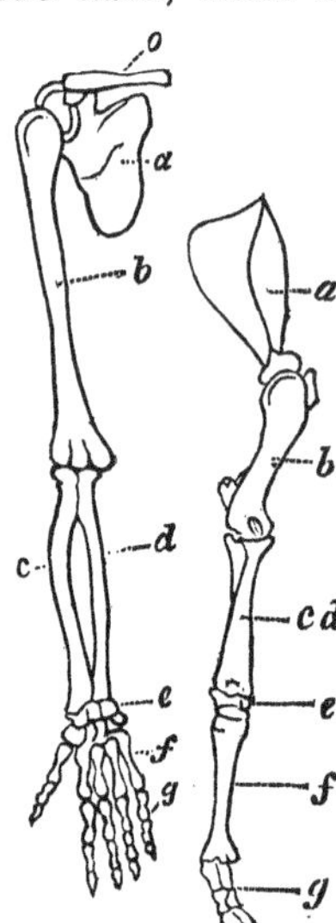

Fig. 34. Fig. 35.

173. The following outlines, in which corresponding bones are indicated by the same letters, will give an idea of the modifications which these bones present in different classes. In the arm of man, (Fig. 34,) the shoulder-blade is flat and triangular; the bone of the arm is cylindrical, and enlarged at its extremities; the bones of the fore-arm are somewhat shorter than the humerus, but more slender; the hand is composed of the following pieces, namely, eight small bones of the carpus, arranged in two rows, five metacarpal bones, which are elongated, and succeed those of the wrist; five fingers of unequal length, one of which, the thumb, is opposed to the four others.

174. In the stag, (Fig. 35,) the bones of the fore-arm are rather longer than that of the arm, and the radius no longer turns upon the ulna, but is blended with it; the metacarpal, or cannon bone, is greatly developed; and, being quite as long as the fore-arm, it is apt to be mistaken for it. The fingers are reduced to two, each of which is surrounded by a hoof, at its extremity.

175. In the arm of the lion, (Fig. 36,) the arm bone is

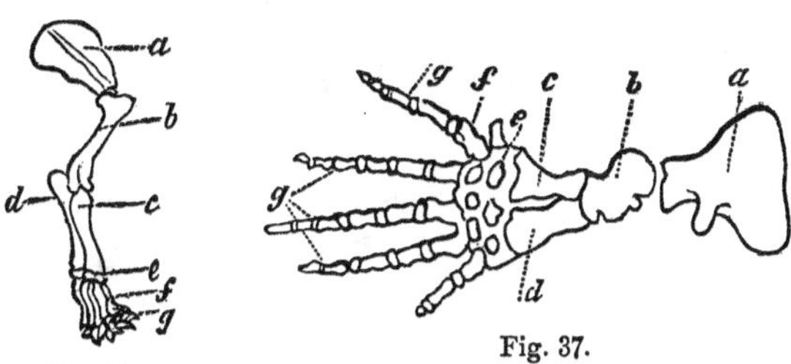

Fig. 36.

Fig. 37.

stouter, the carpal bones are less numerous, and the fingers are short, and armed with strong, retractile claws. In the whale, (Fig. 37,) the bones of the arm and fore-arm are much shortened, and very massive; the hand is broad, the fingers strong, and distant from each other.

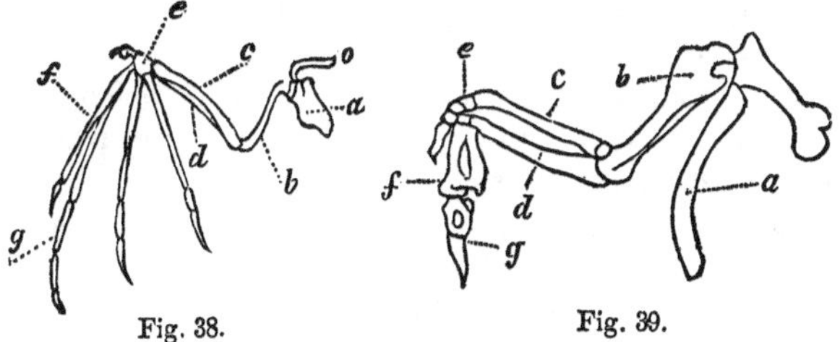

Fig. 38.

Fig. 39.

In the bat, the thumb, which is represented by a small hook, is entirely free, (Fig. 38;) but the fingers are elongated in a disproportionate manner, and the skin is stretched across them, so as to serve the purpose of a wing. In birds, the pigeon for example, (Fig. 39,) there are but two fingers, which are soldered together, and destitute of nails; and the thumb is rudimentary.

176. The arm of the turtle (Fig. 40) is peculiar in having,

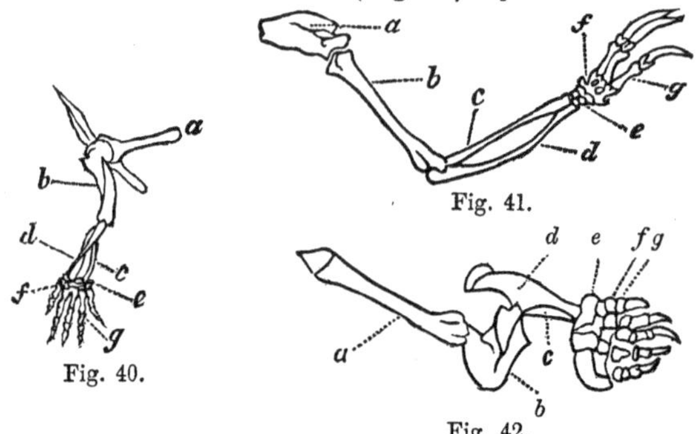

Fig. 40.

Fig. 41.

Fig. 42.

besides the shoulder-blade, two clavicles; the arm-bone is twisted outwards, as well as the bones of the fore-arm, so that the elbow, instead of being behind, is turned forwards; the fingers are long, and widely separated. In the Sloth, (Fig. 41,) the bones of the arm and fore-arm are very greatly elongated, and at the same time very slender; the hand is likewise very long, and the fingers are terminated by enormous non-retractile nails. The arm of the Mole (Fig. 42) is still more extraordinary. The shoulder-blade, which is usually a broad and flat bone, becomes very narrow; the arm-bone, on the contrary, is contracted so much as to seem nearly square; the elbow projects backwards, and the hand is excessively large and stout.

177. In fishes, the form and arrangement of the bones is so peculiar, that it is often difficult to trace their correspondence to all the parts found in other animals; nevertheless, the bones of the fore-arm are readily recognized. In the Cod

(Fig. 43) there are two flat and broad bones, one of which, the ulna, (*d*,) presents a long point, anteriorly. The bones of

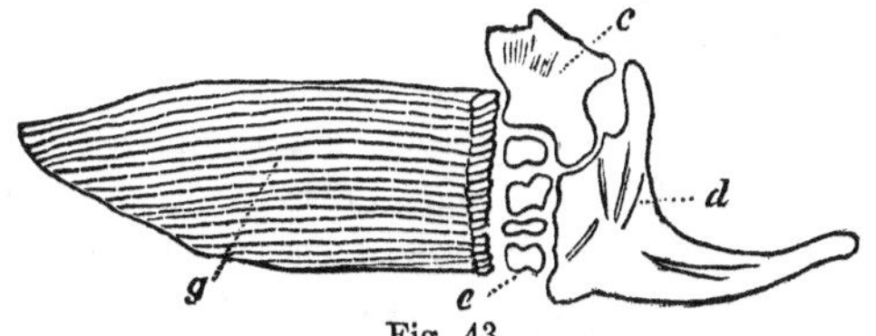

Fig. 43.

the carpus are represented by four nearly square little bones. But in these again there are considerable variations in different fishes, and in some genera they are much more irregular in form. The fingers are but imperfectly represented by the rays of the fin, (*g*,) which are composed of an infinitude of minute bones, articulated with each other. As to the humerus and shoulder, their analogies are variously interpreted by different anatomists.

178. The form of the members is so admirably adapted to the special offices which they are designed to perform, that by a single inspection of the bones of the arm, as represented in the preceding sketches, one might infer the uses to which they are to be put. The arm of man, with its radius turning upon its ulna, the delicate and pliable fingers, and the thumb opposed to them, bespeak an organ for the purpose of handling. The slender and long arm of the sloth, with his monstrous claws, would be extremely inconvenient for walking on the ground, but appropriate for seizing upon the branches of the trees, on which these animals live. The short fingers, armed with retractile nails, indicate the lion, at first glance, to be a carnivorous animal. The arm of the stag, with his very long cannon-bone, and that of the horse, also, with its solitary finger enveloped in a hoof, are organs especially adapted for running. The very slender and greatly elongated fingers of the bat are admirably con-

trived for the spread of a wing, without increasing the weight of the body. The more firm and solid arm of the bird indicates a more sustained flight. The short arm of the whale, with his spreading fingers, resembles a strong oar. The enormous hand of the mole, with its long elbow, is constructed for the difficult and prolonged efforts requisite in burrowing. The twisted arm of the tortoise can be applied to no other movement than creeping. And finally, the arm of the fish, completely enveloped in the mass of the flesh, presents, externally, a mere delicate balancer, the pectoral fin.

179. The posterior members are identical in their structure with the anterior ones. The bones of which they are composed, are, 1. The *pelvis,* (Fig. 46,) which corresponds to the shoulder blade ; 2. The thigh bone, or *femur*, which is a single bone, like the humerus ; 3. The bones of the leg, the *tibia* and *fibula*, which, like the radius and ulna, sometimes coalesce into one bone ; and lastly, the bones of the foot, which are divided, like those of the hand, into three parts, the tarsus or ankle, the metatarsus or instep, and the toes. The modifications are generally less marked than in the arm, inasmuch as there is less diversity of function ; for in all animals, without exception, the posterior extremities are used exclusively for support or locomotion.

180. The anterior extremity of the vertebrates, however varied in form, whether it be an arm, a wing, or a fin, is thus shown to be composed of essentially the same parts, and constructed upon the same general plan. This affinity does not extend to the invertebrates ; for although in many instances their limbs bear a certain resemblance to those of the vertebrates, and are even used for similar purposes, yet they have no real

Fig. 44. Fig. 45.

affinity. Thus the leg of an insect, (Fig. 44,) and that of a lizard, (Fig. 45;) the wing of a butterfly and the wing of a bat, are quite similar in form, position, and use; but in the bat and the lizard, the organ has an internal bony support, which is a part of the skeleton; while the leg of the insect has merely a horny covering, proceeding from one of the rings of the body, and the wing of the butterfly is merely a fold of the skin, showing that the limbs of the Articulata are constructed upon a different plan, (157.) It is by ascertaining and regarding these real affinities, or the fundamental differences, existing between similar organs, that the true natural grouping of animals is to be attained.

2. *Of Standing, and the Modes of Progression.*

181. Standing, or the natural attitude of an animal, depends on the form and functions of the limbs. Most of the terrestrial mammals, and the reptiles, both of which employ all four limbs in walking, have the back-bone horizontal, and resting at the same time upon both the anterior and posterior extremities. Birds, whose anterior limbs are intended for a purpose very different from the posterior, stand upon the latter, when at rest, although the back-bone is still very nearly horizontal. Man alone is designed to stand upright, with his head supported on the summit of the vertebral column. Some monkeys can rise upon the hind legs into the erect posture; but it is evidently a constrained one, and not their habitual attitude.

182. That an animal may stand, it is requisite that the limbs should be so disposed that the centre of gravity, in other words, the point about which the body balances itself, should fall within the space included by the feet. If the centre of gravity is outside of these limits, the animal falls to the side to which the centre of gravity inclines. On this account, the albatross, and some other aquatic birds

which have the feet placed very far back, cannot use them for walking.

183. The more numerous and the more widely separated are the points of support, the firmer an animal stands. On this account, quadrupeds are less liable .to lose their balance than birds. If an animal has four legs, it is not necessary that they should have a broad base. Thus we see that most quadrupeds have slender legs, touching the earth by only a small surface. Broad feet would interfere with each other, and only increase the weight of the limbs, without adding to their stability. Birds are furnished with long toes, which, as they spread out, subserve the purpose of tripods. Moreover, the muscles of the toes are so disposed that the weight of the bird causes them to grasp firmly; hence it is enabled to sleep standing in perfect security upon the roost, without effort.

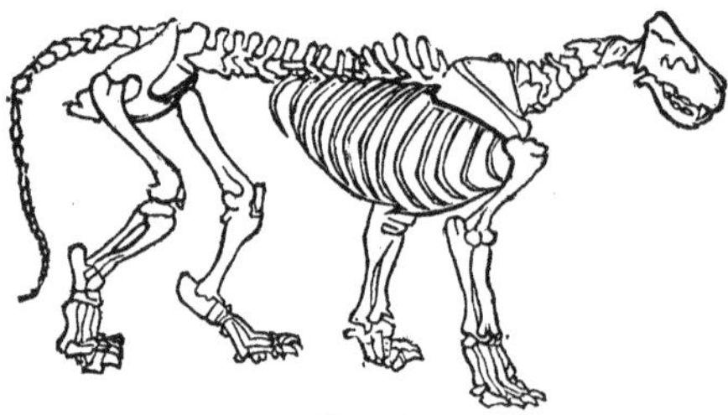

Fig. 46.

184. In quadrupeds, the joints at the junction of the limbs with the body bend freely in only one direction, that is, towards the centre of gravity; so that if one limb yields, the tendency to fall is counteracted by the resistance of the limbs at the other extremity of the body. The same antagonism is observed in the joints of the separate limbs, which are flexed alternately in opposite directions. Thus the thigh bends forwards, and the leg backwards; while the arm bends

backwards, and the fore-arm forwards. Different terms have been employed to express the various modes of progression, according to the rapidity or the succession in which the limbs are advanced.

185. Progression is a forward movement of the body, effected by successively bending and extending the limbs. Walking is the ordinary and natural gait, and other paces are only occasionally employed. When walking is accomplished by two limbs only, as in man, the body is inclined forwards, carrying the centre of gravity in that direction; and while one leg sustains the body, the other is thrown forwards to prevent it from falling, and to sustain it in turn. For this reason, walking has been defined to be a continual falling forwards, continually interrupted by the projection of the legs.

186. The throwing forwards of the leg, which would require a very considerable effort, were the muscles obliged to sustain the weight of the limbs also, is facilitated by a very peculiar arrangement; that is, the joints are perfectly closed up; so that the external pressure of the atmosphere is sufficient of itself to maintain the limbs in place, without the assistance of the muscles. This may be proved by experiment. If we cut away all the muscles around the hip joint, the thigh bone still adheres firmly to the pelvis, but separates the moment a hole is pierced, so as to admit air into the socket.

187. In ordinary walking, the advancing leg touches the ground just before the other is raised; so that there is a moment when the body rests on both limbs. It is only when the speed is very much accelerated, that the two actions become simultaneous. The walking of quadrupeds is a similar process, but with this difference, that the body always rests on at least two legs. The limbs are raised in a determinate order, usually in such a manner that the hind-leg of one side succeeds the fore-leg of the opposite side. Some

animals, as the giraffe, the lama, and the bear, raise both legs of one side at the same moment. This is called *ambling*, or *pacing*.

188. RUNNING consists in the same succession of motions as walking, so accelerated that there is a moment between two steps when none of the limbs touch the ground. In the horse and dog, and in most mammals, a distinction is made between the walk, the trot, the canter, and the gallop, all of which have different positions or measures. The *trot* has but two measures. The animal raises a leg on each side, in a cross direction, that is to say, the right fore-leg with the left hind-leg, and so on. The *canter* has three measures. After advancing the two fore-legs, one after the other, the animal raises and brings forward the two hind-legs, simultaneously. When this movement is greatly urged, there are but two measures; the fore-limbs are raised together as well as the hind-legs; it is then termed a *gallop*.

189. LEAPING consists in a bending of all the limbs, followed by a sudden extension of them, which throws the body forwards with so much force as to raise it from the ground, for an instant, to strike again at a certain distance in advance. For this purpose, the animal always crouches before leaping. Most animals make only an occasional use of this mode of progression, when some obstacle is to be surmounted; but in a few instances, this is the habitual mode. As the hind-legs are especially used in leaping, we observe that all leaping animals have the posterior members very much more robust than the anterior, as the frog, the kangaroo, jerboa, and even the hare. Leaping is also common among certain birds, especially among the sparrows, the thrushes, &c. Finally, there is also a large number of leaping insects, such as the flea, the large tribe of grasshoppers and crickets, in which we find that pair of legs with which leaping is accomplished much more developed than the others.

190. CLIMBING is merely walking upon an inclined or even upright surface. It is usually accomplished by means of sharp nails; and hence many carnivorous animals climb with great facility, such as the cat tribe, the lizards; and many birds, the woodpecker, for instance. Others employ their arms for this purpose, like the bears when they climb a tree; or their hands, and even their tails, like the monkeys; or their beaks, like the parrots. Lastly, there are some whose natural mode of progression is climbing. Such are the sloths, with their arms so long, that, when placed upon the ground, they move very awkwardly; and yet their structure is by no means defective, for in their accustomed movements upon trees they can use their limbs with very great adroitness.

191. Most quadrupeds can both walk, trot, gallop, and leap; birds walk and leap; lizards neither leap nor gallop, but only walk and run, and some of them with great rapidity. No insect either trots or gallops, but many of them leap. Yet their leaping is not always the effect of the muscular force of their legs, as with the flea and grasshopper; but some of them leap by means of a spring, in the form of a hook, attached to the tail, which they bend beneath the body, and which, when let loose, propels them to a great distance, as in the Podurellæ. Still others leap by means of a spring, attached beneath the breast, which strikes against the abdomen when the body is bent; as the spring-beetles, (Elaters.)

192. FLIGHT is accomplished by the simultaneous action of the two anterior limbs, the wings, as leaping is by that of the two hinder limbs. The wings being expanded, strike and compress the air, which thus becomes a support, for the moment, upon which the bird is sustained. But as this support very soon yields, owing to the slight density of the air, it follows that the bird must make the greater and more

rapid efforts to compensate for this disadvantage. Hence it requires a much greater expenditure of strength to fly than to walk; and, therefore, we find the great mass of muscles in birds concentrated about the breast, (Fig. 30.) To facilitate its progress, the bird, after each flap of the wings, brings them against the body, so as to present as little surface as possible to the air; for a still further diminution of resistance, all birds have the anterior part of the body very slender. Their flight would be much more difficult if they had large heads and short necks.

193. Some quadrupeds, such as the flying-squirrel and Galeopithecus, have a fold of the skin at the sides, which may be extended by the legs, and which enables them to leap from branch to branch with more security. But this is not flight, properly speaking, since none of the peculiar operations of flight are performed. There are also some fishes, whose pectoral fins are so extended as to enable them to dart from the water, and sustain themselves for a considerable time in the air; and hence they are called flying-fish. But this is not truly flight.

194. Swimming is the mode of locomotion employed by the greater part of the aquatic animals. Most animals which live in the water swim with more or less facility. Swimming has this in common with flight, that the medium in which it is performed, the water, becomes also the support, and readily yields also to the impulse of the fins. Only, as water is much more dense than air, and as the body of most aquatic animals is of very nearly the same specific gravity as water, it follows that, in swimming, very little effort is requisite to keep the body from sinking. The whole power of the muscles is consequently employed in progression, and hence swimming requires vastly less muscular force than flying.

195. Swimming is accomplished by means of various organs designated under the general term, *fins*, although in an anatomical point of view these may represent very different parts. In the Whales, the anterior extremities and the tail are transformed into fins. In Fishes, the pectoral fins, which represent the arms, and the ventral fins, which represent the legs, are employed for swimming, but they are not the principal organs; for it is by the tail, or caudal fin, that progression is principally effected. Hence the progression of the fish is precisely that of a boat under the sole guidance of the sculling-oar. In the same manner as a succession of strokes alternately right and left propels the boat straight forwards, so the fish advances by striking alternately right and left. To advance obliquely, it has only to strike a little more strongly in the direction opposite to that which he wishes to take. The Whales, on the contrary, swim by striking the water up and down; and it is the same with a few fishes also, such as the rays and the soles. The air-bladder facilitates the rising and sinking of the fish, by enabling it to vary the specific weight of the body.

196. Most land animals swim with more or less ease, by simply employing the ordinary motions of walking or leaping. Those which frequent the water, like the beaver, or which feed on marine animals, as the otter and duck, have webbed feet; that is to say, the fingers are united by a membrane, which, when expanded, acts as a paddle.

197. There is also a large number of invertebrate animals in which swimming is the principal or the only mode of progression. Lobsters swim by means of their tail, and, like the Whales, strike the water up and down. Other crustacea have a pair of legs fashioned like oars; as the posterior legs in sea-crabs, for example. Many insects, likewise, swim with their legs, which are abundantly fringed with hairs to give them surface; as the little water boatmen, (Gyrinus, Dytiscus,) whose mazy dances on the summer streams every one must have observed. The cuttle-fish uses its long ten-

tacles as oars, (Fig. 47;) and some star-fishes (Comatula, Euryale) use their arms with great adroitness, (Fig. 151.) Finally, there are some insects which have their limbs constructed for running on the surface of water, as the water-spiders, (Ranatra, Hydrometra.)

Fig. 47.

198. A large number of animals have the faculty of moving both in the air and on land, as is the case with most birds, and a great proportion of insects. Others move with equal facility, and by the same members, on land and in water, as some of the aquatic birds and most of the reptiles, which latter have even received the name Amphibia, on this account. There are some which both walk, fly, and swim, as the ducks and water-hens; but they do not excel in either mode of progression.

199. However different the movements and offices performed by the limbs may appear to us, according to the element in which they act, we see that they are none the less the effect of the same mechanism. The contraction of the same set of muscles causes the leg of the stag to bend for leaping, the wing of the bird to flap in the air, the arm of the mole to excavate the earth, and the fin of the whale to strike the water.

CHAPTER SIXTH.

NUTRITION.

200. The second class of the functions of animals are those which relate to the maintenance of life and the perpetuation of the species; the functions of *vegetative life*, (59.)

201. The increase of the volume of the body must require additional materials. There is also an incessant waste of particles which, having become unfit for further use, are carried out of the system. Every contraction of a muscle expands the energy of some particles, whose place must be supplied. These supplies are derived from every natural source, the animal, vegetable, and even the mineral kingdoms; and are received under every variety of solid, liquid, and gaseous form. Thus, there is a perpetual interchange of substance between the animal body and the world around. The conversion of these supplies into a suitable material, its distribution to all parts, and the appropriation of it to the growth and sustenance of the body, is called Nutrition in the widest sense of that term.

202. In early life, during the period of growth, the amount of substances appropriated is greater than that which is lost At a later period, when growth is completed, an equilibrium between the matters received and those rejected is established. At a still later period, the equilibrium is again disturbed, more is rejected than is retained, decrepitude begins, and at last the organism becomes exhausted, the functions cease, and *death* ensues.

203. The solids and fluids taken into the body as food are

subjected to a process called *Digestion*, by which the solid portions are reduced to a fluid state also, the nutritive separated from the excrementitious, and the whole prepared to become blood, bone, muscle, &c. The residue is afterwards expelled, together with those particles of the body which require to be renewed, and those which have been derived from the blood by several processes, termed *Secretions.* Matters in a gaseous form are also received and expelled with the air we breathe, by a process called *Respiration.* The nutritive fluids are conveyed to every part of the body by currents, usually confined in vessels, and which, as they return, bring back the particles which are to be either renovated or expelled. This circuit is what is termed the *Circulation.* The function of Nutrition, therefore, combines several distinct processes.

SECTION I.

OF DIGESTION.

204. Digestion, or the process by which the nutritive parts of food are elaborated and prepared to become part of the body, is effected in certain cavities, the *stomach* and *intestines*, or *alimentary canal.* This canal is more or less complicated in the various classes of animals; but there is no animal, however low its organization, without it, in some form, (54.)

Fig. 48.

205. In the polypi, the digestive apparatus is limited to a single cavity. In the Sea Anemone, (Actinia,) for example, it is a pouch, (Fig. 48, *b*,) suspended in

the interior of the body. When the fooc has been sufficiently digested there, it passes, by imbibition, into the general cavity of the body, (*c*,) which is filled with water, and mingling with it, flows thence into all parts of the animal. The jelly-fishes, (Medusæ,) and some Worms, have a distinct stomach, with appendages branching off in every direction, (Fig. 31,) in which a more complete elaboration takes place. The little worms known by the name of Planaria, present a striking example of these ramifications of the intestine, (Fig. 49, *e.*) But here, likewise, the product of digestion mingles with the fluids of the cavity of the body which surround the intestine (*d*) and its branches, and circulation is not yet distinct from digestion.

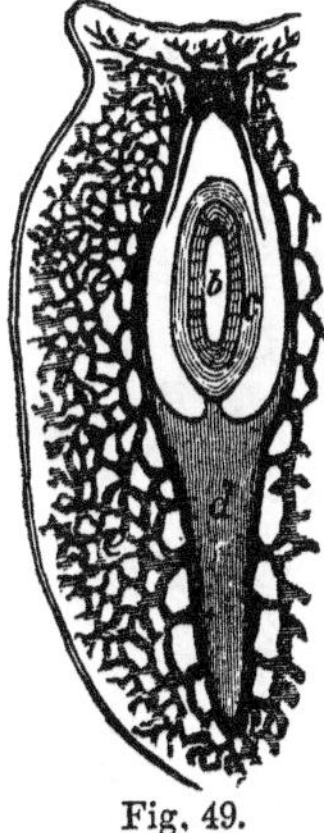
Fig. 49.

206. As we rise in the scale of animals, the functions concerned in nutrition become more and more distinct from each other. Digestion and circulation, no longer confounded, are accomplished separately, in distinct cavities. The most important organs concerned in digestion are the *stomach*, and the *small* and *large intestine.* The first indications of such a distinction are perceived in the higher Radiata, such as the sea-urchins, (Fig. 50,) in which the stomach (*s*) is broader than either extremity of the intestine. The dimensions and form of the cavities of the intestine vary considerably, according to the mode of life of the animal; but the special functions assigned to them are invariable; and the three principal cavities succeed each other, in

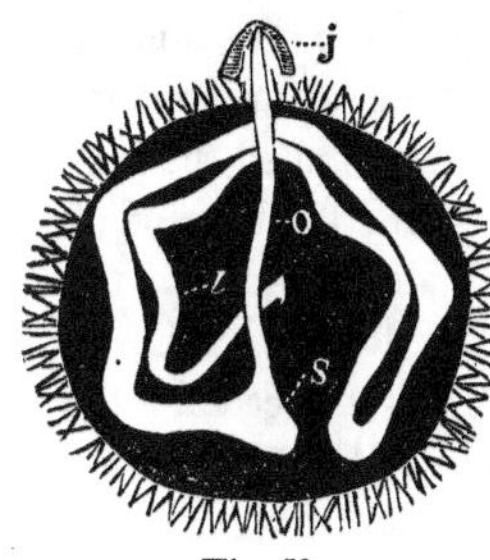

Fig. 50.

every animal where they are found, in an invariable order; first, the stomach, (*s*,) then the intestine, which is small at first, but often enlarged towards its termination. This arrangement may be seen by the following diagrams from a beetle and a land mollusk, where the same letters indicate corresponding parts, (Figs. 51, 52.)

207. From the mouth, (*m*,) the food passes into the stomach through a narrow tube in the neck, called the *œsophagus* or *gullet*, (*o*.) This is not always a direct passage of uniform size; but there is sometimes a pouch, the *crop*, (*c*,) into which the food is first introduced, and which sometimes acquires considerable dimensions, especially in birds, and in some insects and mollusks, (Fig. 51.) In the stomach, the true digestive process is begun. The food no sooner arrives there than changes commence, under the influence of a peculiar fluid called the *gastric juice*, which is secreted by glands lining the interior of the stomach. The digestive action is sometimes aided by the movements of the stomach itself, which, by its strong contractions, triturates the food. This is especially the case in the gizzard of some birds, which, in the hens and ducks, for instance, is a powerful muscular organ. In some of the Crustacea and Mollusks, as the Lobster and Aplysia, there are even solid organs for breaking down the food within the stomach itself.

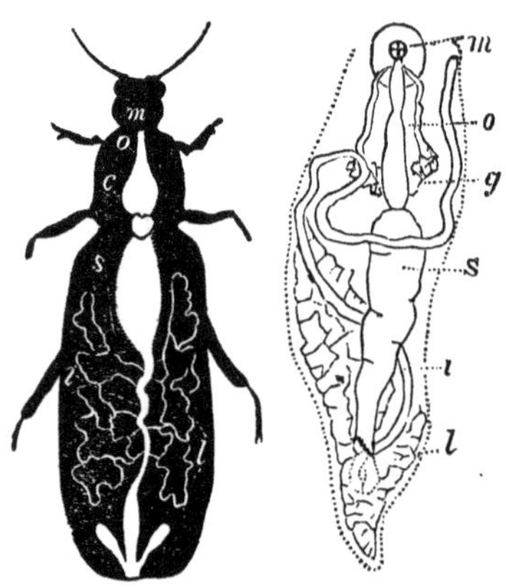

Fig. 51. Fig. 52.

208. The result of this process is the reduction of the food

to a pulpy fluid called *chyme*, which varies in its nature with the food. Hence the function of the stomach has been named *chymification*. With this, the function of digestion is complete in many of the lower animals, and chyme is circulated throughout the body; this is the case in Polypi and Jelly-fishes, and some Worms and Mollusks. In other animals, however, the chyme thus formed is transferred to the intestine, by a peculiar movement, like that of a worm in creeping, which has accordingly received the name of *vermicular* or *peristaltic motion*.

209. The form of the small intestine (*i*) is less variable than that of the stomach. It is a narrow tube, with thin walls, coiled in various directions in the vertebrate animals, but more simple in the invertebrates, especially the insects. Its length varies, according to the nature of the food, being in general longer in herbivorous than in carnivorous animals. In this portion of the canal, the aliment undergoes its complete elaboration, through the agency of certain juices which here mingle with the chyme, such as the bile secreted by the liver, and the pancreatic juice, secreted by the pancreas. The result of this elaboration is to produce a complete separation of the truly nutritious parts, in the form of a milky liquid called *chyle*. The process is called *chylification;* and there are great numbers of animals, such as the Insects, Crabs, and Lobsters, some Worms, and most of the Mollusks, in which the product of digestion is not further modified by respiration, but circulates throughout the body as chyle.

210. The chyle is composed of minute, colorless globules, of a somewhat flattened form, (Fig. 53.) In the higher animals, the Vertebrates, it is taken up and carried into the blood by means of very minute vessels, called *lymphatic vessels* or *lacteals*, which are distributed every where in the walls of the intestine, and communicate

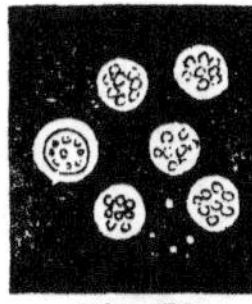

Fig. 53.

with the veins, forming also in their course several glandular masses, as seen in a portion of intestine connected with a vein in Fig. 54; and it is not until thus taken up and mingled with the circulating blood, that any of our food really becomes a part of the living body. Thus freed of the nutritive portion of the food, the residue of the product of digestion passes on to the large intestine, from whence it is expelled in the form of excrement.

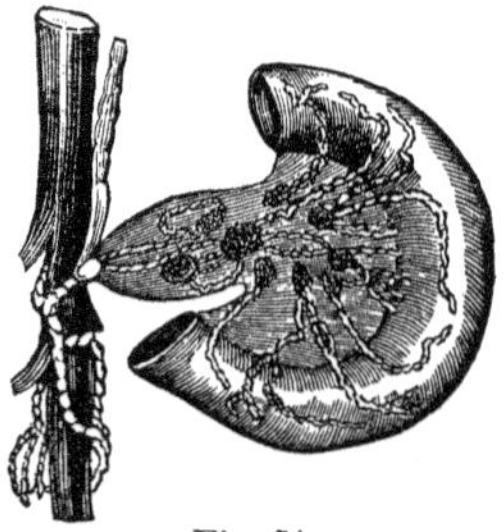
Fig. 54.

211. The organs above described constitute the most essential for the process of digestion, and are found more or less developed in all but some of the radiated animals; but there are, in the higher animals, several additional ones for aiding in the reduction of the food to chyme and chyle, which render their digestive apparatus quite complicated. In the first place, hard parts, of a horny or bony texture, are usually placed about the mouth of those animals that feed on solid substances, which serve for cutting or bruising the food into small fragments before it is swallowed; and, in many of the lower animals, these organs are the only hard portions of the body. This process of subdividing or chewing the food is termed *mastication.*

212. Beginning with the Radiata, we find the apparatus for mastication partaking of the star-like arrangment which characterizes those animals. Thus in Scutella, (Fig. 55,) we have a pentagon composed of five triangular jaws, converging at their summits towards a central aperture which corresponds to the mouth, each one bearing a cutting plate or tooth, like a knife-blade, fitted by one edge into a cleft. The five jaws move towards the centre, and pierce or cut the objects which come between them. In some of the sea-urchins

(Echinus,) this apparatus, which has been called Aristotle's

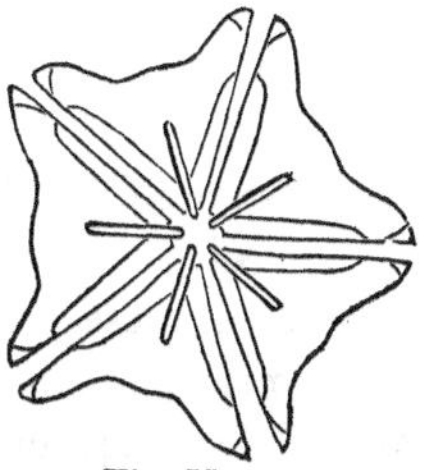
Fig. 55.

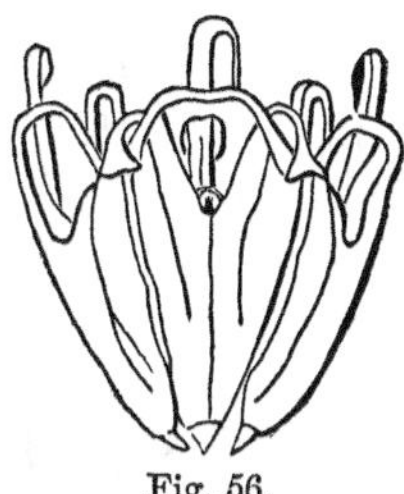
Fig. 56.

lantern, (Fig. 56,) consists of numerous pieces, and is much more complicated. Still, the five fundamental pieces or jaws, each of them bearing a tooth at its point, may be recognized, as in Scutella; only instead of being placed horizontally, they form an inverted pyramid.

213. Among the Mollusks, a few, like the cuttle-fishes, have solid jaws or beaks closely resembling the beak of a parrot, (Fig. 57,) which move up and down as in birds. But a much larger number rasp their food by means of a flat blade coiled up like a watch-spring, the surface of which is covered with innumerable minute tooth-like points of a horny consistence, as seen in a highly magnified portion of the so-called tongue of Natica, (Fig. 58, *a*,) which, however, is only a modification of the beaks of cuttle-fishes.

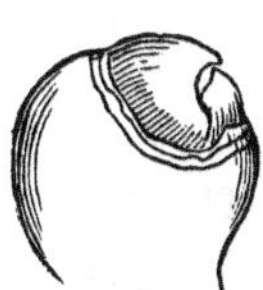
Fig. 57.

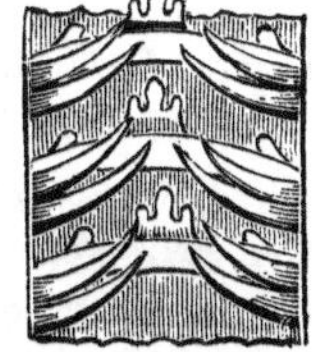
Fig. 58.

214. The Articulata are remarkable, as a class, for the diversity and complication of their apparatus for taking and dividing their food. In some marine worms, Nereis, for example, the jaws consist of a pair of curved, horny instruments, lodged in a sheath, (Fig. 59.) In spiders, these jaws are external, and

Fig. 59.

sometimes mounted on long, jointed stems. Insects which *masticate* their food have, for the most part, at least two pairs of horny jaws, (Figs. 60, 61, *m*, *j*,) besides several additional pieces which serve for seizing and holding their food. Those which live on the fluids which they extract either from plants or from other animals, have the masticatory organs transformed into a trunk or tube for that purpose. This trunk is sometimes rolled up in a spiral manner, as in the butterfly, (Fig. 64;) or it is stiff, and folded beneath the

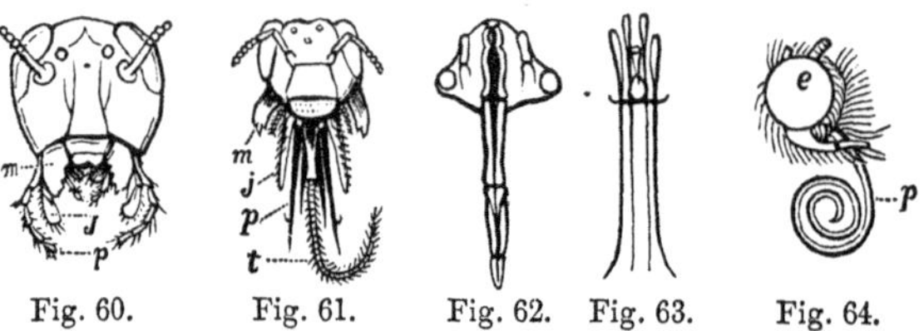

Fig. 60. Fig. 61. Fig. 62. Fig. 63. Fig. 64.

chest, as in the squash-bugs, (Fig. 62,) containing several piercers of extreme delicacy, (Fig. 63,) adapted to penetrate the skin of animals or other objects whose juices they extract; or they are prolonged so as to shield the tongue when thrust out in search of food, as in the bees, (Fig. 61, *t*.) The crabs have their anterior feet transformed into a kind of jaws, and several other pairs of articulated appendages performing ex-

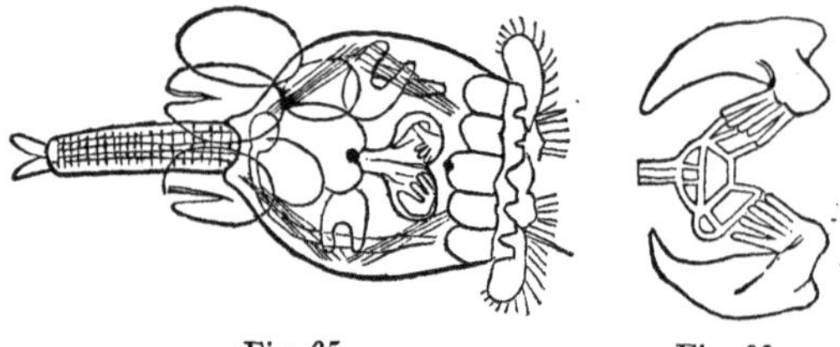
Fig. 65. Fig. 66.

clusively masticatory functions. Even in the microscopic Rotifers, we find very complicated jaws, as seen in a Brachionus, (Fig. 65,) and still more magnified in Fig. 66. But amidst this diversity of apparatus, there is one thing which characterizes all the Articulata, namely, the jaws always move sideways; while those of the Vertebrates and Mollusks move up and down, and those of the Radiata concentrically.

215. In the Vertebrates, the jaws form a part of the bony skeleton. In most of them the lower jaw only is movable, and is brought up against the upper jaw by means of very strong muscles, the temporal and masseter muscles, (Fig. 67, *t*, *m*,) which perform the principal motions requisite for seizing and masticating food.

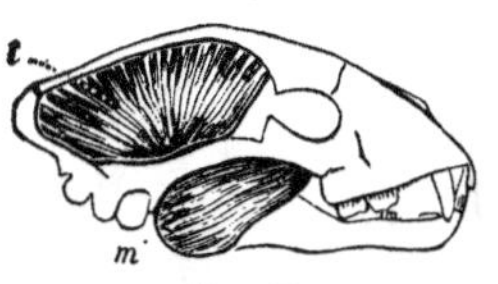

Fig. 67.

216. The jaws are usually armed with solid cutting instruments, the TEETH, or else are enveloped in a horny covering, the *beak*, as in the birds and tortoises, (Fig. 68.) In some of the whales, the true teeth remain concealed in the jaw-bone, and we have instead a range of long, flexible, horny plates or fans, fringed at the margin, which serve as strainers to separate the minute marine animals on which they feed from the water drawn in with them, (Fig. 69.) A few are entirely destitute of teeth, as the ant-eater, (Fig. 70.)

Fig. 68.

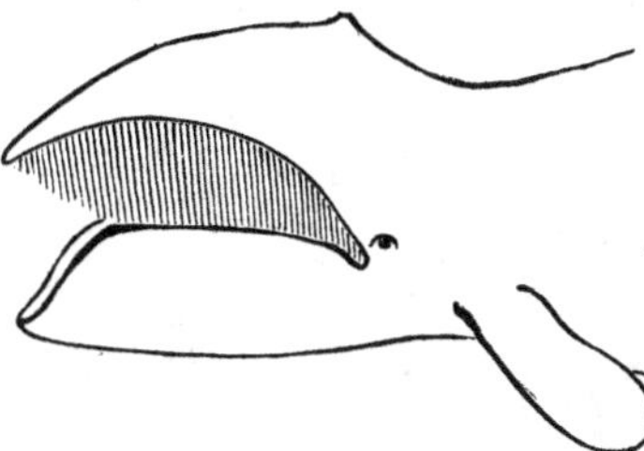
Fig. 69.

217. Though all the vertebrates possess jaws, it must not be inferred that they all chew their food. Many swallow their prey whole; as most birds, tortoises, and whales. Even many of those which are furnished with teeth do not masticate their

food; some using them merely for seizing and securing their prey, as the lizards, frogs, crocodiles, and the great majority of fishes. In such animals, the teeth are nearly all alike in form and structure, as for instance, in the alligator, (Fig. 71,) the porpoises, and many fishes. A few of the latter, some of

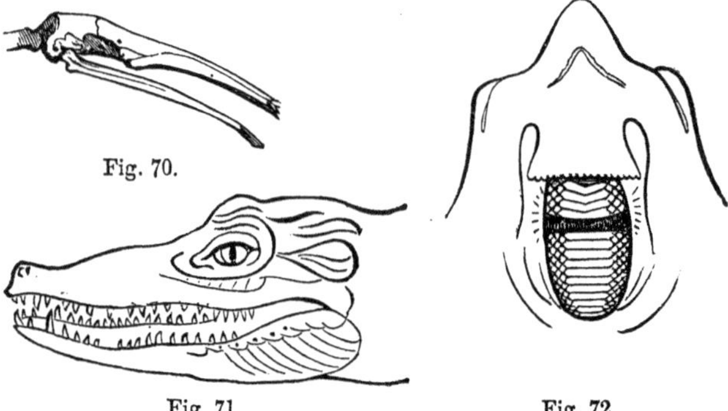

Fig. 70.

Fig. 71.

Fig. 72.

the Rays, for example, have a sort of bony pavement, (Fig. 72,) composed of a peculiar kind of teeth, with which they crush the shells of the mollusks and crabs on which they feed.

218. The Mammals, however, are almost the only vertebrates which can properly be said to masticate their food. Their teeth are well developed, and present great diversity in form, arrangement and mode of insertion. Three kinds of teeth are usually distinguished in most of these animals, whatever may be their mode of life; namely, the cutting teeth, ***incisors***, the

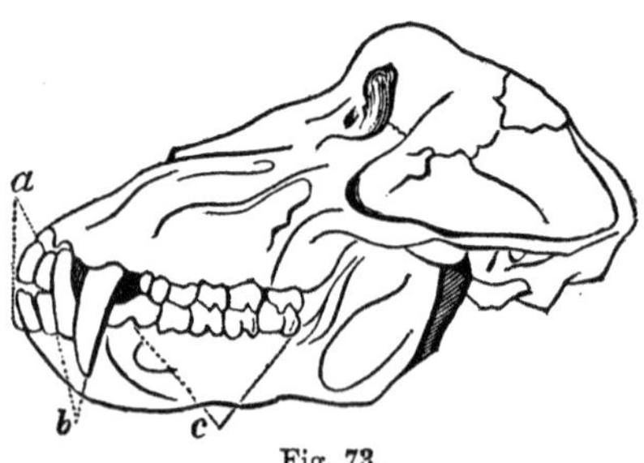

Fig. 73.

tusks or carnivorous teeth, *canines;* and the grinders, *molars,* (Fig. 73.) The *incisors* (*a*) occupy the front of the mouth, the upper ones being set in the intermaxillary bones; they are the most simple and the least varied, have generally a thin cutting summit, and are employed almost exclusively for seizing food, except in the elephant, in which they assume the form of large tusks. The *canines* (*b*) are conical, more elongated than the others, more or less curved, and only two in each jaw. They have but a single root, like the incisors, and in the carnivora become very formidable weapons. In the herbivora, they are wanting, or when existing they are usually so enlarged and modified as also to become powerful organs of offence and defence, although useless for mastication; as in the babyroussa, &c. The *molars* (*c*) are the most important for indicating the habits and internal structure of the animal; they are, at the same time, most varied in shape. Among them we find every transition, from those of a sharp and pointed form, as in the cat tribe, to those with broad and level summits, as in the ruminants and rodents. Still, when most diversified in the same animal, they have one character in common, their roots being never simple, but double or triple, a peculiarity which not only fixes them more firmly, but prevents them from being driven into the jaw in the efforts of mastication.

219. The *harmony of organs* already spoken of (22–24) is illustrated, in a most striking manner, by the study of the teeth of the mammals, and especially of their molar teeth. So constantly do they correspond with the structure of the other parts of the body, that a single molar is sufficient not only to indicate the mode of life of the animal to which i belongs and show whether it feeds on flesh or vegetables, or both, but also to determine the particular group to which it is related. Thus, those beasts of prey which feed on insects, and which on that account have been called Insectivora, such

as the moles and bats, have the molars terminated by several

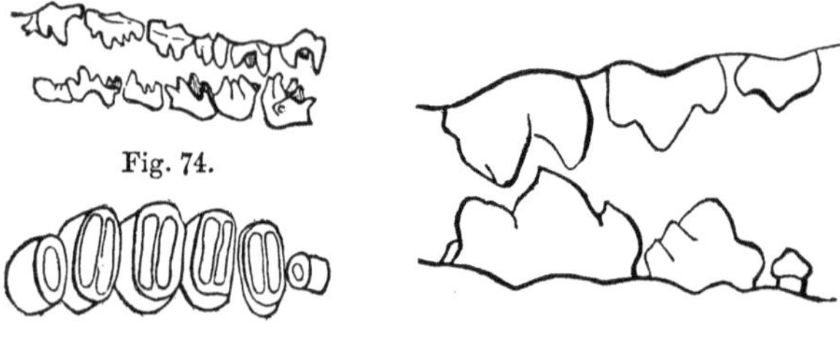

Fig. 74.

Fig. 76. Fig. 75.

sharp, conical points, (Fig. 74,) so arranged that the elevations of one tooth fit exactly into the depressions of the tooth opposite to it. In the true Carnivora, (Fig. 75,) on the contrary, the molars are compressed laterally, so as to have sharp, cutting edges, as in the bats; and they shut by the side of each other, like the blades of scissors, thereby dividing the food with great facility.

220. The same adaptation is observed in the teeth of herbivorous animals. Those which chew the cud, (ruminants,) many of the thick-skinned animals, (pachydermata,) like the elephant, and some of the gnawers, (rodentia,) like the hare, (Fig. 76,) have the summits of the molars flat, like mill-stones, with more or less prominent ridges, for grinding the grass and leaves on which they subsist. Finally, the omnivora, those which feed on both flesh and fruit, like man and the monkeys, have the molars terminating in several rounded tubercles, being thus adapted to the mixed nature of their food.

221. Again, the mode in which the molars are combined with the canines and incisors furnishes excellent means of characterizing families and genera. Even the internal structure of the teeth is so peculiar in each group of animals, and yet subject to such invariable rules, that it is possible to determine with precision the general structure of an animal,

merely by investigating the fragment of a tooth under a microscope.

222. Another process, subsidiary to digestion, is called *insalivation.* Animals which masticate their food have glands, in the neighborhood of the mouth, which secrete a fluid called *saliva.* This fluid mingles with the food as it is chewed, and prepares it also to be more readily swallowed. The salivary glands are generally wanting, or rudimentary, or otherwise modified, in animals which swallow their food without mastication. After it has been masticated and mingled with saliva, it is moved backwards by the tongue, and passes down through the œsophagus, into the stomach. This act is called *deglutition* or *swallowing.*

223. The wisdom and skill of the Creator is strikingly illustrated in the means he has afforded to every creature for securing the means for subsistence. Some animals have no ability to move from place to place, but are fixed to the soil; as the oyster, the polyp, &c. These are dependent for subsistence upon such food as may stray or float near, and they have the means of securing it when it comes within their reach. The oyster closes its shell, and thus entraps its prey; the polyp has flexible arms, (Fig. 77,) capable of great extension, which it throws instantly around any minute animal that comes in contact with it. The cuttle-fish, also, has elongated arms about the mouth, furnished with ranges of suckers, by which it secures its prey, (Fig. 47.)

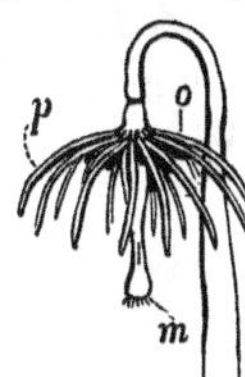

Fig. 77.

224. Some are provided with instruments for extracting food from places which would be otherwise inaccessible. Some of the mollusks, with their rasp-like tongue, (Fig. 58,) perforate the shells of other animals, and thus reach and extract the inhabitant. Insects have various piercers, suckers, or a protractile tongue for the

same purpose, (Figs. 61–64.) Many Annelides, the leeches for example, have a sucker, which enables them to produce a vacuum, and thereby draw out blood from the perforations they make in other animals. Many microscopic animals are provided with hairs or cilia around the mouth, (Fig. 65,) which by their incessant motion produce currents that bring within reach the still more minute creatures or particles on which they feed.

225. Among the Vertebrata, the herbivora generally employ their lips or their tongue, or both together, for seizing the grass or leaves they feed upon. The carnivora use their jaws, teeth, and especially their claws, which are long, sharp even movable, and admirably adapted for the purpose. The woodpeckers have long, bony tongues, barbed at the tip, with which they draw out insects from deep holes and crevices in the bark of trees. Some reptiles also use their tongue to take their prey. Thus, the chameleon obtains flies at a distance of three or four inches, by darting out his tongue, the enlarged end of which is covered with a glutinous substance to which they adhere. The elephant, whose tusks and short neck prevent him from bringing his mouth to the ground, has the nose prolonged into a trunk, which he uses with great dexterity for bringing food and drink to his mouth. Doubtless the mastodon, once so abundant in this country, was furnished with a similar organ. Man and the monkeys employ the hand exclusively, for prehension.

226. Some animals drink by suction, like the ox, others by lapping, like the dog. Birds simply fil. the beak with water, then, raising the head, allow it to run down into the crop. It is difficult to say how far aquatic animals require water with their food ; it seems, however, impossible that they should swallow their prey without introducing at the same time some water into their stomach. Of many among the lowest animals, such as the Polyps it is well

known that they frequently fill the whole cavity of their body with water, through the mouth, the tentacles, and pores upon the sides, and empty it at intervals through the same openings. And thus the aquatic mollusks introduce water into special cavities of the body, or between their tissues, through various openings, while others pump it into their blood vessels, through pores at the surface of their body. This is the case with most fishes.

226 *a*. Besides the more conspicuous organs above described, there are among the lower animals various microscopic apparatus for securing their prey. The lassos of polypi have been already mentioned incidentally, (223.) They are minute cells, each containing a thin thread coiled up in its cavity, which may be thrown out by inversion, and extend to a considerable length beyond the sac to which it is attached. Such lassos are grouped in clusters upon the tentacles, or scattered upon the sides of the Actinia and of most polypi. They occur also in similar clusters upon the tentacles and the disk of jelly-fishes. The nettling sensation produced by the contact of many of these animals is undoubtedly owing to the lasso cells. Upon most of the smaller animals, they act as a sudden, deadly poison. In Echinoderms, such as star-fishes, and sea-urchins, we find other microscopic organs in the form of clasps, placed upon a movable stalk. The clasps, which may open and shut alternately, are composed of serrated or hooked branches, generally three in number, closing concentrically upon each other. With these weapons, star-fishes not more than two inches in diameter may seize and retain shrimps of half that length, notwithstanding their efforts to disentangle themselves.

CHAPTER SEVENTH.

OF THE BLOOD AND CIRCULATION.

227. The nutritive portions of the food are poured into the general mass of fluid which pervades every part of the body, out of which every tissue is originally constructed, and from time to time renewed. This fluid, in the general acceptation of the term, is called blood ; but it differs greatly in its essential constitution in the different groups of the Animal Kingdom. In polypi and medusæ, it is merely chyme, (208 ;) in most mollusks and articulates it is chyle, (209 ;) but in vertebrates it is more highly organized, and constitutes what is properly called Blood.

228. The Blood, when examined by the microscope, is found to consist of a transparent fluid, the *serum*, consisting chiefly of albumen, fibrin, and water, in which float many rounded, somewhat compressed bodies, called blood disks.

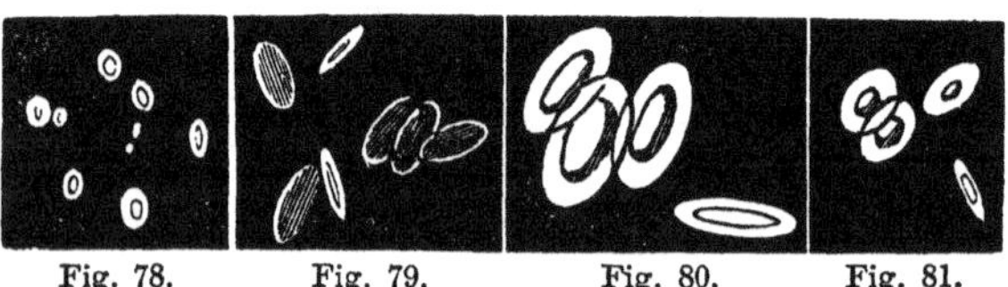

Fig. 78. Fig. 79. Fig. 80. Fig. 81.

These vary in number with the natural heat of the animal from which the blood is taken. Thus, they are more numerous in birds than in mammals, and more abundant in the latter than in fishes. In man and other mammals they are very small and nearly circular, (Fig. 78;) they are somewhat larger, and of an oval form, in birds and fishes, (Figs. 79, 81;) and still larger in reptiles, (Fig. 80.)

229. The color of the blood in the vertebrates is bright red; but in some invertebrates, as the crabs and mollusks, the nutritive fluid is nearly or quite colorless; while in the worms and some echinoderms, it is variously colored yellow, orange, red, violet, lilac, and even green.

230. The presence of this fluid in every part of the body is one of the essential conditions of animal life. A perpetual current flows from the digestive organs towards the remotest parts of the surface; and such portions as are not required for nutriment and secretions return to the centre of circulation, mingled with fluids which need to be assimilated to the blood, and with particles of the body which are to be expelled, or, before returning to the heart, are distributed in the liver. The blood is kept in an incessant CIRCULATION for this purpose.

231. In the lowest animals, such as the polypi, the nutritive fluid is simply the product of digestion (chyme) mingled with water in the common cavity of the viscera, with which it comes in immediate contact, as well as with the whole interior of the body. In the jelly-fishes, which occupy a somewhat higher rank, a similar liquid is distributed by prolongations of the principal cavity to different parts of the body, (Fig. 31.) Currents are produced in these, partly by the general movements of the animal, and partly by means of the incessant vibrations of microscopic fringes, called *vibratile cilia*, which overspread the interior. In most of the mollusks and articulates, the blood (chyle) is also in immediate contact with the viscera, water being mixed with it in mollusks; the vessels, if there are any, not forming a complete circuit, but emptying into various cavities which interrupt their course.

232. In animals of still higher organization, as the vertebrates, we find the vital fluid enclosed in an appropriate set of vessels, by which it is successively conveyed throughout the system to supply nutriment and secretions, and to the respiratory organs, where it absorbs oxygen, or, in other words, becomes *oxygenated.*

233. The vessels in which the blood circulates are of two kinds: 1. The *arteries*, of a firm, elastic structure, which may be distended or contracted, according to the volume of their contents, and which convey the blood from the centre towards the surface, distributing it to every point of the body. 2. The *veins*, of a thin, membranous structure, furnished within with valves, (Fig. 82, *v*,) which aid in sustaining the column of blood, only allowing it to flow from the periphery towards the centre. The arteries constantly subdivide into smaller and smaller branches; while the veins commence in minute twigs, and are gathered into branches and larger trunks, to unite finally into a few stems, near the centre of circulation.

Fig. 82.

234. The extremities of the arteries and veins are connected by a net-work of extremely delicate vessels, called *capillary vessels*, (Fig. 83.) They pervade every portion of the body, so that almost no point can be pricked without drawing blood. Their office is to distribute the nutritive fluid to the organic cells, where all the important processes of nutrition are performed, such as the alimentation and growth of all organs and tissues, the elaboration of bile, milk, saliva, and

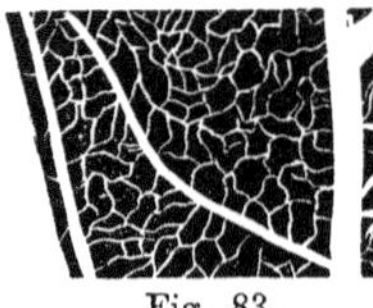
Fig. 83.

other important products derived from blood, the removal of effete particles and the substitution of new ones, and all those changes by which the bright blood of the arteries becomes the dark blood of the veins ; and again, in the cells of the respiratory organs which the capillaries supply, the dark venous blood is oxygenated and restored to the bright scarlet hue of the arterial blood.

235. Where there are blood-vessels in the lowest animals, the blood is kept in motion by the occasional contraction of some of the principal vessels, as in the worms. Insects have a large vessel running along the back, furnished with valves, so arranged that, when the vessel contracts, the blood can flow only towards the head, and, being thence distributed to the body, is returned again into the *dorsal vessel*, (Fig. 84,) by fissures at its sides.

Fig. 84.

236. In all the higher animals there is a central organ, the *heart*, which forces the blood through the arteries towards the periphery, and receives it again on its return. The HEART is a hollow, muscular organ, of a conical form, which dilates and contracts at regular intervals, independently of the will. It is either a single cavity, or is divided by walls into two, three, or four compartments, as seen in the following diagrams. These modifications are important in their connection with the respiratory organs, and indicate the higher or lower rank of an animal, as determined by the quality of the blood distributed in those organs.

237. In the mammals and birds the heart is divided by a vertical partition into two cavities, each of which is again divided into two compartments, one above the other, as seen in the diagram, (Fig. 85.) The two upper cavities are called

auricles, and the two lower *ventricles*. Reptiles have two

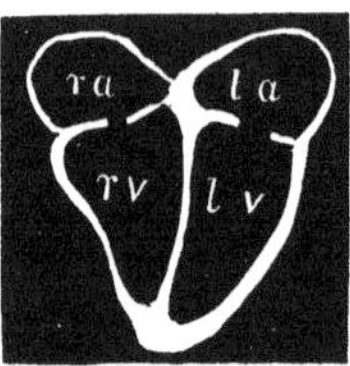

Fig. 85.

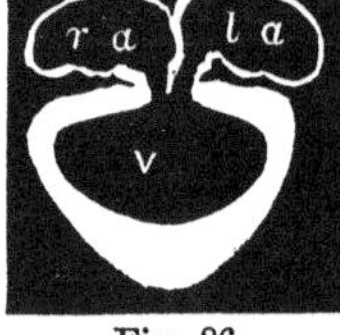

Fig. 86.

Fig. 87.

auricles and one ventricle, (Fig. 86.) Fishes have one auricle and one ventricle only, (Fig. 87.)

238. The auricles do not communicate with each other, in adult animals, nor do the ventricles. The former receive the blood from the body and the respiratory organs, through veins, and each auricle sends it into the ventricle beneath, through an opening guarded by a valve, to prevent its reflux; while the ventricles, by their contractions, force the blood through arteries into the lungs, and through the body generally.

239. The two auricles dilate at the same instant, and also contract simultaneously; so also do the ventricles. These successive contractions and dilatations constitute the pulsations of the heart. The contraction is called *systole*, and the dilatation is called *diastole*. Each pulsation consists of two movements, the diastole or dilatation of the ventricles, during which the auricles contract, and the systole or contraction of the ventricles, while the auricles dilate. The frequency of the pulse varies in different animals, and even in the same animal, according to its age, sex, and the degree of health. In adult man, they are commonly about seventy beats per minute.

240. The course of the blood in those animals which have four cavities to the heart is as follows, beginning with the left ventricle, (Fig. 85, *l. v.*) By the contraction of this

ventricle, the blood is driven through the main arterial trunk, called the *aorta*, (Fig. 90, *a*,) and is distributed by its branches throughout the body; it is then collected by the veins, carried back to the heart, and poured into the right auricle, (Fig. 85, *r a*,) which sends it into the right ventricle *r v*.) The right ventricle propels it through another set of arteries, the *pulmonary arteries*, (Fig. 90, *p*,) to the lungs, (*l*;) it is there collected by the *pulmonary veins*, and conveyed to the left auricle, (Fig. 85, *l a*,) by which it is returned to the left ventricle, thus completing the circuit.

241. Hence the blood in performing its whole circuit passes twice through the heart. The first part of this circuit, the passage of the blood through the body, is called the *great circulation;* and the second part, the passage of the blood through the lungs, is the *lesser* or *pulmonary circulation:* this double circuit is said to be a *complete* circulation. In this case the heart may be justly regarded as two hearts conjoined, and in fact the whole of the lesser circulation intervenes in the passage of the blood from one side of the heart to the other; except that during the embryonic period there is an opening between the two auricles, which closes as soon as respiration commences.

242. In reptiles, (Fig. 86,) the venous blood from the body is received into one auricle, and the oxygenated blood from the lungs into the other. These throw their contents into the single ventricle below, which propels the mixture in part to the body, and in part to the lungs; but as only the smaller portion of the whole quantity is sent to the lungs in a single circuit, the circulation is said to be *incomplete.* In the Crocodiles, the ventricle has a partition which keeps separate the two kinds of blood received from the auricles; but the mixture soon takes place by means of a special artery, which passes from the pulmonary artery to the aorta.

243. In fishes, (Fig. 87,) the blood is carried directly

from the ventricle to the gills, which are their chief respiratory organs; thence it passes into arteries for distribution to the system in general, and returns by the veins to the auricle. Here the blood, in its circuit, passes but once through the heart; but the heart of a fish corresponds nevertheless to the heart of a mammal, and not to one half of it, as has often been maintained, for the gills are not lungs.

244. Crabs and other crustacea have but a single ventricle, without an auricle. In the mollusks, there is likewise but a single ventricle, as in Natica, (Fig. 88, *h.*) Some have in addition one or two auricles. These auricles are sometimes so disjoined as to form so many isolated hearts, as in the cuttle-fish. Among Radiata, the sea-urchins are provided with a tubular heart.

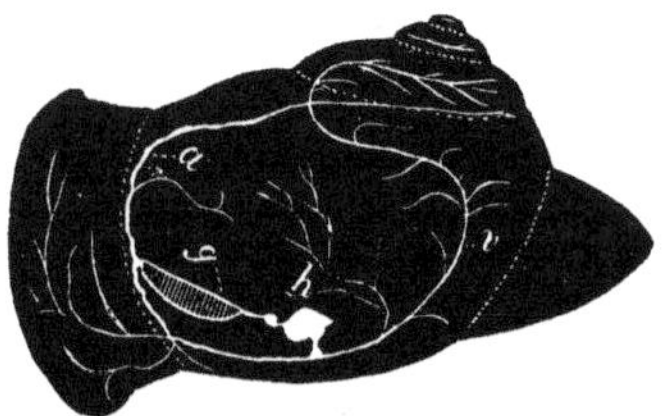

Fig. 88.

CHAPTER EIGHTH.

OF RESPIRATION.

245. For the maintenance of its vital properties, the blood must be submitted to the influence of the air. This is true of all animals, whether they live in the atmosphere or in the water. No animal can survive for any considerable period of time without air; and the higher animals almost instantly die when deprived of it. It is the office of RESPIRATION to bring the blood into communication with the air.

246. Among animals which breathe in the open air, some have a series of tubes branching through the interior of the body, called *tracheæ*, (Fig. 89, *t*,) opening externally upon the sides of the body, by small apertures, called *stigmata*, (*s*;) as in insects and in some spiders. But the most common mode of respiration is by means of LUNGS, a pair of peculiar spongy or cellular organs, in the form of large pouches, which are the more complicated in proportion to the quantity of air to be consumed.

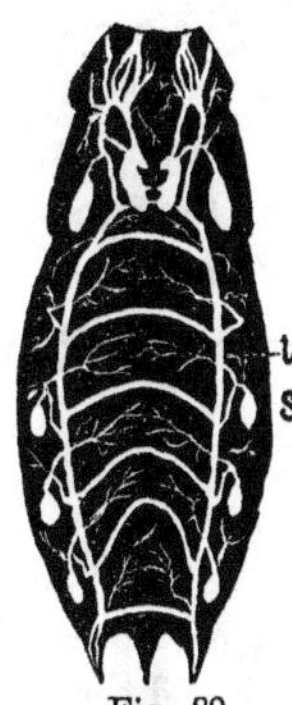

Fig. 89.

247. In the lower vertebrata, provided with lungs, they form a single organ; but in the higher classes they are in pairs, placed in the cavity formed by the ribs, one on each side of

the vertebral column, and enclosing the heart (*h*) between them, (Fig. 90, *l l*.) The lungs communicate with the atmosphere by means of a tube composed of cartilaginous rings which arises from the back part of the mouth, and divides below, first into a branch for each organ, and then into innumerable branches penetrating their whole mass, and finally terminating in minute sacs. This tube is the *trachea* or *windpipe*, (*w*,) and its branches are the *bronchi*. In the higher air-breathing animals the lungs and heart occupy an apartment by themselves, the *chest*, which is separated from the other contents of the lower arch of the vertebral column, (161,) by a fleshy partition, called the *diaphragm*, passing across the cavity of the body, and arching up into the chest. The only access to this apartment from without is by the glottis, (Fig. 22, *o*,) through the trachea.

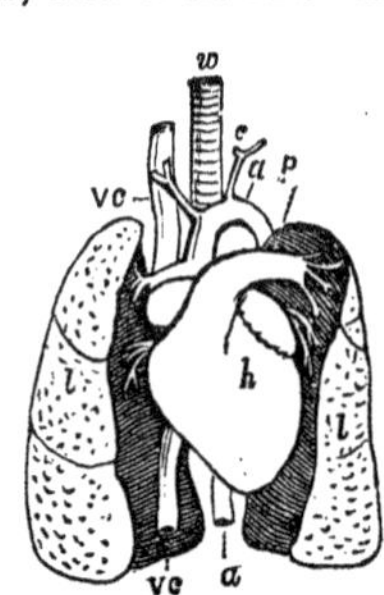

Fig. 90.

248. The mechanism of respiration by lungs may be compared to the action of a bellows. The cavity of the chest is enlarged by raising the ribs, the arches of which naturally slope somewhat downward, but more especially by the contraction of the diaphragm, whereby its intrusion into the chest is diminished. This enlargment causes the air to rush in through the trachea, distending the lung so as to fill the additional space. When the diaphragm is again relaxed, and the ribs are allowed to subside, the cavity is again diminished, and the air expelled. These movements are termed *inspiration* or *inhalation*, and *expiration*. The spongy pulmonary substance being thus distended by air, the blood sen. from the heart is brought into such contact with it as to allow the requisite interchange to take place, (235.)

249. The respiration of animals breathing in water is ac-

complished by a different apparatus. The air is to be derived from the water, in which more or less is always diffused. The organs for this purpose are called *branchiæ* or *gills*, and are either delicate tufts or plumes floating outside of the body, as in some of the marine worms, (Fig. 33,) and many mollusks, (Fig. 91, *g*;) or they consist of delicate combs and brushes, as in fishes, (Fig. 92,) crabs, and most mollusks, (Fig. 88, *g*.) These gills are always so situated that the water has free access to them. In the lower aquatic animals, such as the polypi, and some jelly-fishes and mollusks, respiration takes place by the incessant motions of vibratory cilia, which fringe both the outside and the cavities of the body; the currents they produce bringing constantly fresh supplies of water, containing air, into contact with the respiratory surface.

Fig. 91.

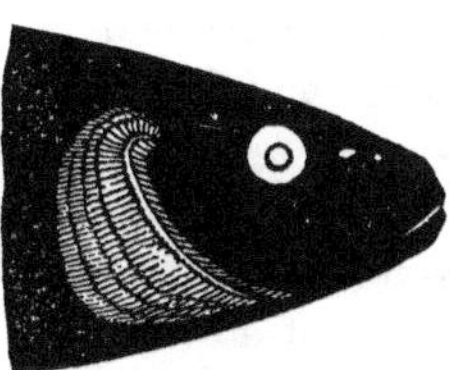
Fig. 92.

250. Many animals living in water, however, rise to the surface and breathe the atmosphere there, or are furnished with the means of carrying away a temporary supply of air, whilst others are furnished with reservoirs in which the blood requiring oxygenation may be accumulated, and their stay under water prolonged. This is the case with the seals, whales, tortoises, frogs, many insects and mollusks, &c.

251. The vivifying power of the air upon the blood is due to its oxygen. If an animal be confined for a time in a closed vessel, and the contained air be afterwards examined, a considerable portion of its oxygen will have disappeared, and another gas of a very different character, namely, carbonic acid gas, will have taken its place. The essential office of respiration is to supply oxygen to the blood, at the same time that carbon is removed from it.

252. An immediately obvious effect of respiration in the red-blooded animals is a change of color; the blood, in passing through the respiratory organs, being changed from a very dark purple to a bright scarlet. In the great circulation (241) the scarlet blood occupies the arteries, and is usually called *red blood*, in contradistinction from the venous blood, which is called *black blood*. In the lesser circulation, on the contrary, the arteries carry the dark, and the veins the red blood.

253. The quantity of oxygen consumed by various animals in a given time has been accurately ascertained by experiment. It has been found, for instance, that a common-sized man consumes, on an average, about 150 cubic feet in twenty-four hours; and as the oxygen constitutes but 21 per cent. of the atmosphere, it follows that he inhales, during a day, about 700 cubic feet of atmospheric air. In birds, the respiration is still more active, while in reptiles and fishes it is much more sluggish.

254. The energy and activity of an animal is, therefore somewhat dependent on the activity of its respiration. Thus the toad, whose movements are very sluggish, respires much more slowly than the mammals, birds, and even insects; and it has been ascertained that a butterfly, notwithstanding its comparatively diminutive size, consumes more oxygen than a toad.

255. The circulation and respiration have a reciprocal influence upon each other. If the heart be powerful, or if on violent exercise a more rapid supply of blood to repair the consequent waste is demanded, (201,) respiration must be proportionally accelerated to supply air to the greater amount of blood sent to the lungs. Hence the panting occasioned by running or other unusual efforts of the muscles. On the other hand, if respiration be hurried, the blood is rendered more stimulating by greater oxygenation, and causes an ac-

celeration of the circulation. The quantity of air consumed varies, therefore, with the proportion of the blood which is sent to the lungs.

256. The proper temperature of an animal, or what is termed ANIMAL HEAT, depends on the combined activity of the respiratory and circulating systems, and is in direct proportion to it. In many animals the heat is maintained at a uniform temperature, whatever may be the variations of the surrounding medium. Thus, birds maintain a temperature of about 108° Fahrenheit; and in a large proportion of mammals it is generally from 95° to 105°. These bear the general designation of *warm-blooded animals.*

257. Reptiles, fishes, and most of the still lower animals, have not this power of maintaining a uniform temperature. The heat of their body is always as low as from 35° to 50°, but varies perceptibly with the surrounding medium, being often, however, a little above it when the external temperature is very low, though some may be frozen without the loss of life. For this reason, they are denominated *cold-blooded animals;* and all animals which have such a structure of the heart that only a part of the blood which enters it is sent to the respiratory organs, are among them, (243.)

258. The production of animal heat is obviously connected with the respiratory process. The oxygen of the respired air is diminished, and carbonic acid takes its place. The carbonic acid is formed in the body by the combination of the oxygen of the air with the carbon of the blood. The chemical combination attending this function is, therefore, essentially the same as that of combustion. It is thus easy to understand how the natural heat of an animal is greater, in proportion as respiration is more active. How far nutrition in general, and more particularly *assimilation*, by which the liquid parts are fixed and solidified, is connected with the maintenance of the proper temperature of animals, and the

uniform distribution of heat through the body, has not yet been satisfactorily ascertained.

259. Some of the higher warm-blooded animals do not maintain their elevated temperature during the whole year; but pass the winter in a sort of lethargy called HIBERNATION, or the *hibernating sleep*. The marmot, the bear, the bat, the crocodile, and most reptiles, furnish examples. During this state the animal takes no food; and as it respires only after very prolonged intervals, its heat is diminished, and its vital functions generally are much reduced. The structural cause of hibernation is not ascertained; but the phenomena attending it fully illustrate the laws already stated, (254–8.)

260. There is another point of view in which respiration should be considered, namely, with reference to the buoyancy of animals, or their power of rising in the atmosphere, and their ability to live at different depths in the water, under a diminished or increased pressure. The organs of respiration of birds and insects are remarkably adapted for the purpose of admitting at will a greater quantity of air into their body, the birds being provided with large pouches extending from the lungs into the abdominal cavity and into the bones of the wing. In insects the whole body is penetrated by air tubes, the ramifications of their tracheæ, which are enlarged at intervals into wider cells; whilst most of the aquatic animals are provided with minute, almost microscopic tubes, penetrating from the surface into the substance, or the cavities of the body, admitting water into the interior, by which they thus adapt their whole system to pressures which would otherwise crush them. These tubes may with propriety be called water-tubes. In fishes, they penetrate through the bones of the head and shoulder, through skin and scales, and communicate with the blood vessels and heart, into which they pour water; in mollusks they are more numerous in the fleshy parts, as, for example, in the

foot, which they help to distend, and communicate with the main cavity of the body, supplying it also with liquid; in echinoderms they pass through the skin, and even through

---

260 *a*. In order fully to appreciate the homologies between the various respiratory apparatus observed in different animals, it is necessary to resort to a strict comparison of the fundamental connections of these organs with the whole system of organization, rather than to the consideration of their special adaptation to the elements in which they live. In Vertebrates, for instance, there are two sets of distinct respiratory organs, more or less developed at different periods of life, or in different groups. All Vertebrates, at first, have gills arising from the sides of the head, and directly supplied with blood from the heart; but these gills are the essential organs of respiration only in fishes and some reptiles, and gradually disappear in the higher reptiles, as well as in birds and Mammalia, towards the close of their embryonic growth. Again, all Vertebrates have lungs, opening in or near the head; but the lungs are fully developed only in Mammalia, birds, and the higher reptiles, in proportion as the branchial respiration is reduced; whilst in fishes the air-bladder constitutes a rudimentary lung.

260 *b*. In Articulates, there are also two sorts of respiratory organs; aerial, called tracheæ in insects, and lungs in spiders; and aquatic, in crustacea and worms, called gills. But these tracheæ and lungs open separately upon the two sides of the body, (air never being admitted through the mouth or nostrils in Articulates;) the gills are placed in pairs; those which are like the tracheæ occupying a similar position, so that there are nearly as many pairs of tracheæ and gills as there are segments in these animals, (Figs. 89 and 33.) The different respiratory organs in Articulates are in reality mere modifications of the same apparatus, as their mode of formation and successive metamorphoses distinctly show, and cannot be compared with either the lungs or gills of Vertebrates; they are special organs not found in other classes, though they perform the same functions. The same may be said of the gills and lungs of mollusks, which are essentially alike in structure, the lungs of snails and slugs being only a modification of the gills of aquatic mollusks; but these two kinds of organs differ again in their structure and relations from the tracheæ and gills of Articulates, as much as from the lungs and gills

the hard shell, whilst in polyps they perforate the walls of the general cavity of the body, which they constantly fill with water.

---

of Vertebrates. In those Radiates which are provided with distinct respiratory organs, such as the Echinoderms, we find still another typical structure, their gills forming bunches of fringes around the mouth, or rows of minute vesicles along the radiating segments of the body.

CHAPTER NINTH.

OF THE SECRETIONS.

261. WHILE, by the process of digestion, a homogeneous fluid is prepared from the food, and supplies new material to the blood, another process is also going on, by which the blood is analyzed, as it were; some of its constituents being selected and so combined as to form products for useful purposes, while other portions of it which have become useless or injurious to the system are taken up by different organs, and expelled in different forms. This process is termed SECRETION.

262. The organs by which these operations are performed are much varied, consisting either of flat surfaces or membranes, of minute simple sacs, or of delicate elongated tubes, all lined with minute cells, called *epithelium cells*, which latter are the real agents in the process. Every surface of the body is covered by them, and they either discharge their products directly upon the surface, as on the mucous membrane, or they unite in clusters and empty into a common duct, and discharge by a single orifice, as is the case with some of the intestinal glands, and of those from which the perspiration issues upon the skin, (Fig. 94.)

Fig. 93.

263. In the higher animals, where separate organs for special purposes are multiplied, numerous sacs and tubes are assembled into compact masses, called *glands*. Some of these are of large size, such as the salivary glands, the kidneys, and the liver. In these, clusters of sacs open into a common canal, and this canal unites with similar ones forming larger trunks, such as we find in the salivary glands, (Fig. 93,) and finally they all discharge by a single duct.

264. By the organs of secretion, two somewhat different purposes are effected, namely, fluids of a peculiar character are selected from the blood, for important uses, such as the saliva, tears, milk, &c., some of which differ but little in their composition from that of the blood itself, and might be retained in the blood with impunity; or, the fluids selected are such as are positively injurious, and cannot remain in the blood without soon destroying life. These atter are usually termed EXCRETIONS.

265. As the weight of the body, except during its period of active growth, remains nearly uniform, it follows that it must daily lose as much as it receives; in other words, the excretions must equal in amount the food and drink taken, with the exception of the small proportion discharged by the alimentary canal. Some of the most important of these outlets will be now indicated.

266. We have already seen (37) that all animal tissues admit of being traversed by liquids and gases. This mutual transmission of fluids from one side of a membrane to the other is termed *endosmosis* and *exosmosis*, or imbibition and transudation, and is a mechanical, rather than a vital, phenomenon, inasmuch as it takes place in dead as well as in

living tissues. The bloodvessels, especially the capillaries. share this property. Hence portions of the circulating fluids escape through the walls of the vessels and pass off at the surface. This superficial loss is termed *exhalation.* It is most active where the bloodvessels most abound, and accordingly is very copious from the air-tubes of the lungs and from the skin. The loss in this way is very considerable ; and it has been estimated that, under certain circumstances, the body loses, by exhalation, five eighths of the whole weight of the substances received into it.

267. The skin, or outer envelop of the body, is otherwise largely concerned in the losses of the body. Its layers are constantly renewed by the tissues beneath, and the outer dead layers are thrown off. This removal is sometimes gradual and continual, as in man. In fishes and many mollusks, it comes off in the form of slime, which is, in fact, composed of cells detached from the surface of the skin. Sometimes the loss is periodical, when it is termed *moulting.* Thus, the mammals cast their hair, and the deer their horns, the birds their feathers, the serpents their skins, the crabs their test, the caterpillars their outer envelop, with all the hairs growing from it.

268. The skin presents such a variety of structure in the different groups of animals as to furnish excellent distinctive characters of species, genera, and even families, as will hereafter be shown. In the vertebrates we may recognize several distinct layers, of unequal thickness, as may be seen in figure 94, which represents a magnified section of the human skin, traversed by the sudoriferous canals. The lower and thickest layer, (*a*,) is the *cutis*, or true skin, and is the part which is tanned into leather. Its surface presents numerous papillæ, in which the nerves of general sensation terminate ; they also contain a fine network of bloodvessels,

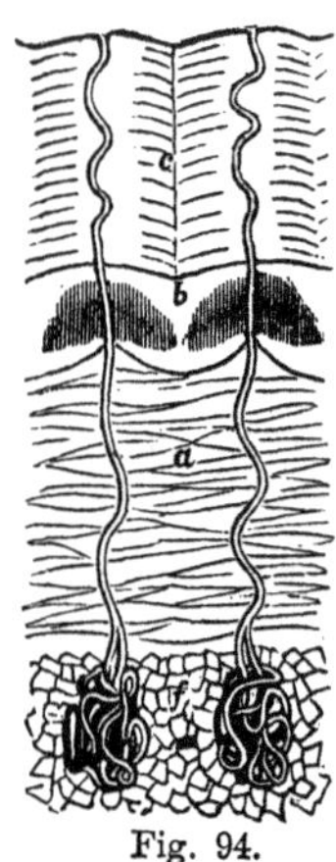

Fig. 94.

usually termed the *vascular layer*. The superficial layer (*c*) is the *epidermis*, or cuticle. The cells of which it is composed are distinct at its inner portion, but become dried and flattened as they are pushed outwards. It is supplied with neither vessels nor nerves, and, consequently, is insensible. Between these two layers, and more especially connected with the cuticle, is the *rete mucosum*, (*b*,) a very thin layer of cells, some of which contain the pigment which gives the complexion to the different races of men and animals. The scales of reptiles, the nails and claws of mammals, and the solid coverings of the Crustacea, are merely modifications of the epidermis. On the other hand, the feathers of birds and the scales of fishes arise from the vascular layer.

269. Of all the Excretions, if we except that from the Lungs, the bile seems to be the most extensive and important; and hence a liver, or some analogous organ, by which bile is secreted, is found in animals of every department; while some, or all, of the other glands are wanting in the lower classes of animals. In Vertebrates, the liver is the largest of all the organs of the body. In mollusks, it is no less preponderant. In the gasteropods, like the snail, it envelopes the intestine in its convolutions, (Fig. 52 ;) and in the acephala, like the clam and oyster, it generally surrounds the stomach. In insects it is found in the shape of long tubes, variously contorted and interlaced, (Fig. 51.) In the Radiata, this organ is largely developed, especially among the echinoderms. In the star-fishes it extends into

all the recesses of the rays; and, in color and structure, resembles the liver of mollusks. Even in polyps, we find peculiar brown cells lining the digestive cavity, which, probably, perform functions similar to those of the liver in the higher animals.

270. The great importance of the respiratory organs in discharging carbon from the blood has already been spoken of, (245, 251.) The substances removed by the liver and the lungs are of the same class, being those which are destitute of nitrogen. These organs seem, in some sense, subsidiary to each other; and hence, in those animals where the respiratory organs are largely developed, the biliary organs are comparatively small, and vice versa. Another and opposite class of impurities, and no less pernicious if retained in the blood, is removed by the KIDNEYS; and, consequently, organs answering to the kidneys are found very far down in the series of animals. Most of the peculiar ingredients of the urine are capable of assuming solid, crystalline forms; and, in some animals, as in reptiles and birds, the whole secretion of the kidneys is solid. In most cases, however, the urinary salts are largely diluted with water; and, as the lungs and liver are supplementary to each other in the removal of carbon, so the lungs, the kidneys, and the skin mutually relieve each other in the removal of the watery portions of the blood.

CHAPTER TENTH.

EMBRYOLOGY.

SECTION I

OF THE EGG.

271. The functions of vegetative life, of which we have treated in the preceding chapters, namely, digestion, circulation, respiration, and secretion, have for their end the preservation of the individual. We have now to treat of the functions that serve for the perpetuation of the species, namely, those of reproduction, (200.)

272. It has been generally admitted that animals as well as plants are the offspring of individuals of the same kind; and *vice versa*, that none of them can give birth to individuals differing from themselves; but recent investigations have modified to a considerable extent this view, as we shall see hereafter.

273. Reproduction in animals is almost universally accomplished by the association of individuals of two kinds, *males* and *females*, living commonly in pairs or in flocks, each of them characterized by peculiarities of structure and external appearance. As this distinction prevails throughout the animal kingdom, it is always necessary, if we would obtain a correct and complete idea of a species, to take into account the peculiarities of both sexes. Every one is familiar with the differences between the cock and the hen, the lion and the lioness, &c. Less prominent peculiarities are observed in

most Vertebrates. Among Articulata, the differences are no less striking, the males being often of a different shape and color, as in crabs, or having even more complete organs, as in many tribes of insects, where the males have wings, while the females are destitute of them, (Fig. 147.) Among mollusks, the females have often a wider shell.

274. Even higher distinctions than specific ones are based upon peculiarities of the sexes; for example, the whole class of Mammalia is characterized by the fact that the female is furnished with organs for nourishing her young with a peculiar liquid, the milk, secreted by herself. Again, the Marsupial, such as the opossum and kangaroo, are distinguished by the circumstance that the female has a pouch into which the young are received in their immature condition at birth.

275. That all animals are produced from eggs, (*Omne vivum ex ovo*,) is an old adage in Zoölogy, which modern researches have fully confirmed. In tracing back the phases of animal life, we invariably arrive at an epoch when the incipient animal is enclosed within an egg. It is then called an *embryo*, and the period passed in this condition is called the *embryonic period.*

276. Before the various classes of the animal kingdom had been attentively studied during the embryonic period, all animals were divided into two great divisions: the *oviparous*, comprising those which lay eggs, such as birds, reptiles, fishes, insects, mollusks, &c., and the *viviparous*, which bring forth their young alive, like the mammalia, and a few from other orders, as the sharks, vipers, &c. This distinction lost much of its importance when it was shown that viviparous animals are produced from eggs, as well as the oviparous; only that their eggs, instead of being laid before the development of the embryo begins, undergo their early changes in the body of the mother. Production from

eggs should, therefore, be considered as a universal characteristic of the Animal Kingdom.

277. *Form of the Egg.* — The general form of the egg is more or less spherical. The eggs of birds have the form of an elongated spheroid, narrow at one end; and this form is so constant, that the term *oval* has been universally adopted to designate it. But this is by no means the usual form of the eggs of other animals. In most instances, on the contrary, they are spherical, especially among the lower animals. Some have singular appendages, as those of the skates and sharks, (Fig. 95,) which are shaped like a hand-barrow, with four hooked horns at the corners. The eggs of the hydra, or fresh water polyp, are thickly covered with prickles, (Fig. 96.) Those of certain insects, the Podurella, for example, are furnished with filaments which give them a hairy aspect, (Fig. 97;) others are cylindrical or prismatic; and frequently the surface is sculptured.

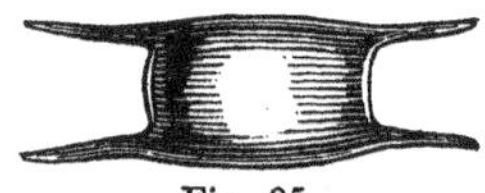

Fig. 95.

Fig. 96.

Fig. 97.

278. *Formation of the Egg.* — The egg originates within peculiar organs, called *ovaries*, which are glandular bodies, usually situated in the abdominal cavity. So long as the eggs remain in the ovary, they are very minute in size. In this condition they are called *ovarian*, or *primitive eggs.* They are identical in all animals, being, in fact, merely little cells (v) containing yolk, (y,) and including other smaller cells, the germinative vesicle, (g,) and the germinative dot, (d.) The yolk itself, with its membrane, (v,) is formed while the egg remains in the ovary. It is afterwards enclosed in another envelope, the shell membrane, which may remain soft, (s,)

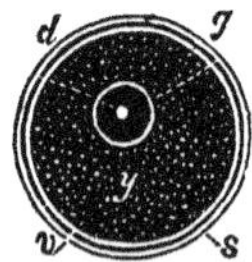

Fig. 98.

or be further surrounded by calcareous deposits, the shell proper, (Fig. 101, *s.*) The number of these eggs is large, in proportion as the animal stands lower in the class to which it belongs. The ovary of a herring contains more than 25,000 eggs; while that of birds contains a much smaller number, perhaps one or two hundred only.

279. *Ovulation.*—Having attained a certain degree of maturity, which varies in different classes, the eggs leave the ovary. This is called *ovulation*, and must not be confounded with the laying of the eggs, which is the subsequent expulsion of them from the abdominal cavity, either immediately, or through a special canal, the *oviduct.* Ovulation takes place at certain seasons of the year, and never before the animal has reached a particular age, which is commonly that of its full growth. In a majority of species, ovulation is repeated for a number of years consecutively, generally in the spring in terrestrial animals, and frequently several times a year; most of the lower aquatic animals, however, lay their eggs in the fall, or during winter. In others, on the contrary, it occurs but once during life, at the period of maturity, and the animal soon afterwards dies. Thus the butterfly and most insects die, shortly after having laid their eggs.

280. The period of ovulation is one of no less interest to the zoölogist than to the physiologist, since the peculiar characteristics of each species are then most clearly marked. Ovulation is to animals what flowering is to plants; and, indeed, few phenomena are more interesting to the student of nature than those exhibited by animals at the pairing season. Then their physiognomy is the most animated, their song the most melodious, and their attire the most brilliant. Some birds appear so different at this time, that zoölogists are always careful to indicate whether or not a bird is represented at the breeding season. Fishes, and many other animals, are ornamented with much brighter colors at this period.

281. *Laying.* — After leaving the ovary, the eggs are either discharged from the animal, that is, *laid;* or they continue their development within the parent animal, as is the case in some fishes and reptiles, as sharks and vipers, which, for that reason, have been named *ovo-viviparous* animals. The eggs of the mammalia are not only developed within the mother, but become intimately united to her; this peculiar mode of development has received the name of *gestation.*

282. Eggs are sometimes laid one by one, as in birds; sometimes collectively and in great numbers, as in the frogs, the fishes, and most of the invertebrates. The queen ant of the African termites lays 80,000 eggs in twenty-four hours; and the common hair-worm, (Gordius,) as many as 8,000,000 in less than one day. In some instances they are united in clusters by a gelatinous envelop; in others they are enclosed in cases or between membranous disks, forming long strings, as in the eggs of the Pyrula shell, (Fig. 99.) The conditions under which the eggs of different animals are placed, on being laid, are very different. The eggs of birds, and of some insects, are deposited in nests constructed for that purpose by the parent. Other animals carry their eggs attached to their bodies; sometimes under the tail, as in the lobsters and crabs, sometimes hanging in large bundles on both sides of the tail, as in the Monoculus, (Fig. 100, *a.*)

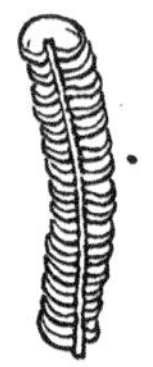

Fig. 99.

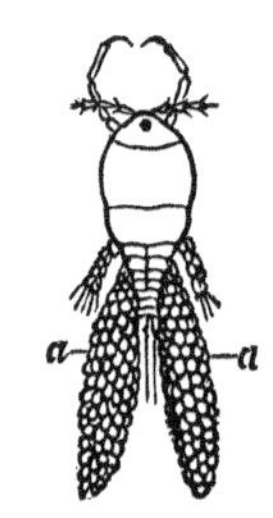

Fig. 100.

283. Some toads carry them on the back, and, what is most extraordinary, it is the male which undertakes this office. Many mollusks, the Unio for example, have them enclosed between the folds of the gills during incubation. In the jelly fishes and polyps, they hang in clusters, either

outside, (Fig. 77, *o*,) or inside, at the bottom of the cavity of the body. Some insects, such as the gad-flies, deposit their eggs on other animals. Finally, many abandon their eggs to the elements, taking no further care of them after they have been laid; such is the case with most fishes, some insects, and many mollusks. As a general rule, it may be said that animals take the more care of their eggs and brood as they occupy a higher rank in their respective classes.

284. The development of the embryo does not always take place immediately after the egg is laid. A considerable time, even, may elapse before it commences. Thus, the first eggs laid by the hen do not begin to develop until the whole number which is to constitute the brood is deposited. The eggs of most butterflies, and of insects in general, are laid in autumn, in temperate climates, and remain unchanged until the following spring. During this time, the principle of life in the egg is not extinct, but is simply inactive, or in a latent state. This tenacity of life is displayed in a still more striking manner in plants. The seeds, which are equivalent to eggs, preserve for years, and even for ages, their power to germinate. Thus, there are some well-authenticated cases in which wheat taken from the ancient catacombs of Egypt has been made to sprout and grow.

285. A certain degree of warmth is requisite for the hatching of eggs. Those of birds, especially, require to be submitted, for a certain length of time, to a uniform temperature, corresponding to the natural heat of the future chicken, which is naturally supplied by the body of the parent. In other words, *incubation* is necessary for their growth. Incubation, however, is not a purely vital phenomenon, but may be easily imitated artificially. Some birds of warm climates dispense with this task; for example, the ostrich often contents herself with depositing her eggs in the sand of the desert, leaving them to be hatched by the sun. In

like manner, the eggs of most birds may be hatched by maintaining them at the proper temperature by artificial means. Some fishes are also known to build nests and to sit upon their eggs, as the sticklebacks, sun-fishes, and cat-fishes; but whether they impart heat to them or not, is doubtful.

Before entering into the details of embryonic transformations, a few words are necessary respecting the composition of the egg.

286. *Composition of the Egg.* — The egg is composed of several substances, varying in structure, as well as in appearance. Thus, in a hen's egg, (Fig. 101,) we have first a calcareous *shell*, (*s*,) lined by a double membrane, the ***shell membrane***, (*m*;) then an albuminous substance, the ***white***, (*a*,) in which several layers may be distinguished; within this we find the *yolk*, (*y*,) enclosed in its membrane; and before it was laid, there was in the midst of the latter a minute vesicle, the *germinative vesicle*, (Fig. 98, *g*,) containing a still smaller one, the *germinative dot*, (*d*.) These different parts are not equally important in a physiological point of view. The most conspicuous of them, namely, the shell and the white, are not essential parts, and therefore are often wanting; while the yolk, the germinative vesicle, and the germinative dot are found in the eggs of all animals; and out of these, and of these only, the germ is formed, in the position shown by Fig. 101, *e*.

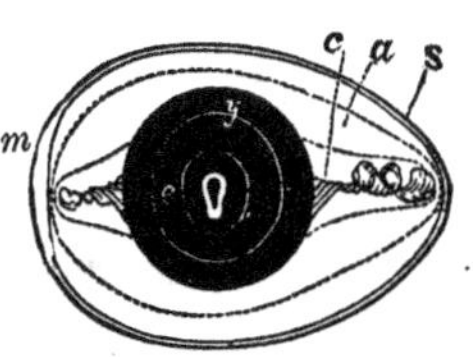

Fig. 101.

287. The *vitellus* or *yolk* (Fig. 101, *y*) is the most essential part of the egg. It is a liquid of variable consistence, sometimes opaque, as in the eggs of birds, sometimes transparent and colorless, as in the eggs of some fishes and mollusks. On examination under the microscope, it appears to be composed of an accumulation of granules and oil-drops.

The yolk is surrounded by a very thin skin, the *vitelline membrane*, (Fig. 98, *v*.) In some insects, when the albumen is wanting, this membrane, surrounded by a layer of peculiar cells, forms the exterior covering of the egg, which, in such cases, is generally of a firm consistence, and sometimes even horny.

288. The *germinative vesicle* (Fig. 98, *g*) is a cell of extreme delicacy, situated, in the young egg, near the middle of the yolk, and easily recognized by the greater transparency of its contents when the yolk is in some degree opaque, as in the hen's egg, or by its outline, when the yolk itself is transparent, as in eggs of fishes and mollusks. It contains one or more little spots, somewhat opaque, appearing as small dots, the *germinal dots*, (*d*.) On closer examination, these dots are themselves found to contain smaller nucleoli.

289. The *albumen*, or white of the egg, (Fig. 101, *a*,) is a viscous substance, generally colorless, but becoming opaque white on coagulation. Voluminous as it is in birds' eggs, it nevertheless plays but a secondary part in the history of their development. It is not formed in the ovary, like the yolk, but is secreted by the oviduct, and deposited around the yolk, during the passage of the egg through that canal. On this account, the eggs of those animals in which the oviduct is wanting, are generally without the albumen. In birds, the albumen consists of several layers, one of which, the *chalazæ*, (*c*,) is twisted. Like the yolk, the albumen is surrounded by a membrane, the *shell membrane*, (*m*,) which is either single or double, and in birds, as also in some reptiles and mollusks, is again protected by a calcareous covering, forming a true shell, (*s*.) In most cases, however, this envelop continues membranous, particularly in the eggs of the mollusks, most crustaceans and fishes, salamanders, frogs, &c. Sometimes it is horny, as in the sharks and skates.

SECTION II.

DEVELOPMENT OF THE YOUNG WITHIN THE EGG.

290. The formation and development of the young animal within the egg is a most mysterious phenomenon. From a hen's egg, for example, surrounded by a shell, and composed, as we have seen, (Fig. 101,) of albumen and yolk, with a minute vesicle in its interior, there is produced, at the end of a certain time, a living animal, composed apparently of elements entirely different from those of the egg, endowed with organs perfectly adapted to the exercise of all the functions of animal and vegetative life, having a pulsating heart, a digestive apparatus, organs of sense for the reception of outward impressions, and having, moreover, the faculty of performing voluntary motions, and of experiencing pain and pleasure. These phenomena are certainly sufficient to excite the curiosity of every intelligent person.

291. By opening eggs which have been subjected to incubation during different periods of time, we may easily satisfy ourselves that these changes are effected gradually. We thus find that those which have undergone but a short incubation exhibit only faint indications of the future animal; while those upon which the hen has been sitting for a longer period include an embryo chicken proportionally more developed. Modern researches have taught us that these gradual changes, although complicated, and at first sight so mysterious, follow a constant law in each great division of the Animal Kingdom.

292. The study of these changes constitutes that peculiar branch of Physiology called EMBRYOLOGY. As there are differences in the four great departments of the Animal

Kingdom perceptible at an early stage of embryonic life, quite as obvious as those found at maturity, and as the phases of embryonic development furnish important indications for the natural classification of animals, we propose to give the outlines of Embryology, so far as it may have reference to Zoölogy.

293. In order to understand the successive steps of embryonic development, we must bear in mind that the whole animal body is formed of tissues, the elements of which are cells, (39.) These cells, however, are more or less diversified and modified, or even completely metamorphosed in the full grown animal; but, at the commencement of embryonic life, the whole embryo is composed of minute cells of nearly the same form and consistence, originating within the yolk, and constantly undergoing changes under the influence of life. New cells are successively formed, while others disappear, or are modified and so transformed as to become bones, muscles, nerves, &c.

294. We may form some idea of this singular process, by noticing how, in the healing of a wound, new substance is supplied by the transformation of blood. Similar changes take place in the embryo, during its early life; only, instead of being limited to some part of the body, they pervade the whole animal.

295. The changes commence, in most animals, soon after the eggs are laid, and are continued without interruption until the development of the young is completed; in others, birds for example, they proceed only to a certain extent, and are then suspended until incubation takes place. The yolk, which at first consists of a mass of uniform appearance, gradually assumes a diversified aspect. Some portions become more opaque, and others more transparent; the germinal vesicle, which was in the midst of the yolk, rises to its upper part where the germ is to be formed. These early changes

are accompanied, in some animals, by a rotation of the yolk within the egg, as may be distinctly seen in some of the mollusks, especially in the snails.

296. At the same time, the yolk undergoes a peculiar process of segmentation. It is first divided into halves, forming distinct spheres, which are again regularly subdivided into two more, and so on, till the whole yolk assumes the appearance of a mulberry, each of the spheres of which it is composed having in its interior a transparent vesicle. This is the case in mammalia, most mollusks, worms, &c. In many animals, however, as in the naked reptiles and fishes,* this segmentation is only partial, the divisions of the yolk not extending across its whole mass.

297. But whether complete or partial, this process leads to the formation of *a germ* comprising the whole yolk, or rising above it as a disk-shaped protuberance, composed of little cells, which has been variously designated under the names of germinative disk, proligerous disk, blastoderma, germinal membrane. In this case, however, that portion of the yolk which has undergone less obvious changes forms, nevertheless, part of the growing germ. The disk again gradually enlarges, until it embraces the whole, or nearly the whole, of the yolk.

298. At this early epoch, namely, a few days, and sometimes a few hours, after development has begun, the germ proper consists of a single layer composed

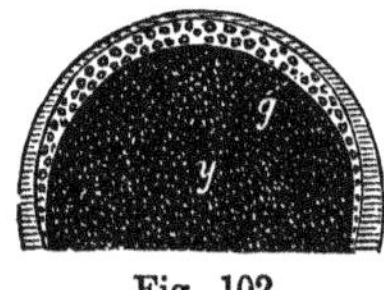

Fig. 102.

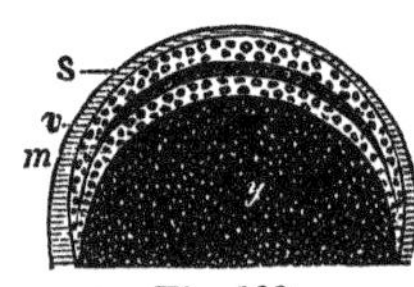

Fig. 103.

* In the Birds and higher reptiles we find, in the mature egg, a peculiar organ, called cicatricula, which may, nevertheless, have been formed by a similar process before it was laid.

of very minute cells, all of which are alike in appearance and form, (Fig. 102, *g*.) But soon after, as the germ increases in thickness, several layers may be discerned, in vertebrated animals, (Fig. 103,) which become more and more distinct.

299. The upper layer, (*s*,) in which are subsequently formed the organs of animal life, namely, the nervous system, the muscles, the skeleton, &c., (59,) has received the name of *serous* or *nervous layer*. The lower layer, (*m*,) which gives origin to the organs of vegetative life, and especially to the intestines, is called the *mucous* or *vegetative layer*, and is generally composed of larger cells than those of the upper or serous layer. Finally, there is a third layer, (*v*,) interposed between the two others, giving rise to the formation of blood and the organs of circulation; whence it has been called *blood layer*, or *vascular layer*.

300. From the manner in which the germ is modified, we can generally distinguish, at a very early epoch, to what department of the animal kingdom an individual is to belong. Thus, in the Articulata, the germ is divided into segments, indicating the transverse divisions of the body, as, for example, in the embryo of the crabs, (Fig. 104.) The germ of the vertebrated animals, on the other hand, displays a longitudinal furrow, which marks the position the future back-bone is to occupy, (Fig. 105.)

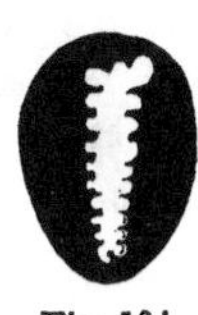

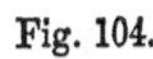

Fig. 104.

Fig. 105.

301. The development of this furrow is highly important, as indicating the plan of structure of vertebrated animals in general, as will be shown by the following figures, which represent vertical sections of the embryo at different epochs.*

* In these figures, the egg is supposed to be cut down through the middle, so that only the cut edge of the embryo is seen; whereas, if viewed

At first the furrow (Fig. 106, *b*) is very shallow, and a lit-

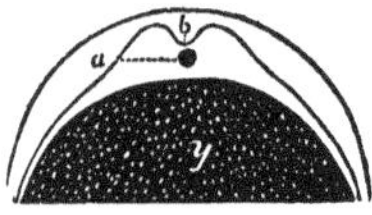
Fig. 106.

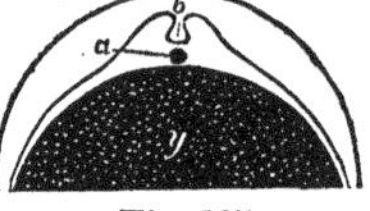
Fig. 107.

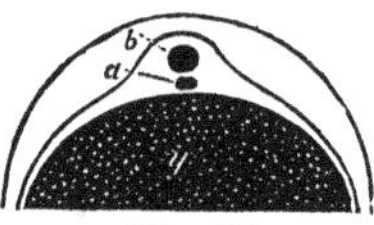
Fig. 108.

tle transparent, narrow band appears under it, called the *primitive stripe*, (*a*.) The walls of the furrow consist of two raised edges formed by a swelling of the germ along both sides of the primitive stripe. Gradually, these walls grow higher, and we perceive that their summits have a tendency to approach each other, as seen in Fig. 107; at last they meet and unite completely, so that the furrow is now changed into a closed canal, (Fig. 108, *b*.) This canal is soon filled with a peculiar liquid, from which the spinal marrow and brain are formed at a later period.

302. The primitive stripe is gradually obliterated by a peculiar organ of a cartilaginous nature, the *dorsal cord*, formed in the lower wall of the dorsal canal. This is found in the embryos of all vertebrates, and is the representative of the back-bone. In the mean time, the margin of the germ gradually extends farther and farther over the yolk, so as finally to enclose it entirely, and form another cavity in which the organs of vegetative life are to be developed. Thus the embryo of vertebrates has two cavities, namely, the upper one, which is very small, containing the nervous system, and the lower, which is much larger, for the intestines, (161.)

303. In all classes of the Animal Kingdom, the embryo proper rests upon the yolk, and covers it like a cap. But the direction by which its edges approach each other, and

---

from above, it would extend over the yolk in every direction, and the furrow at *b*, of Fig. 106, would appear as in Fig. 105.

unite to form the cavity of the body, is very unlike in different animals; and these several modes are of high importance in classification. Among the Vertebrates, the embryo lies with its face or ventral surface towards the yolk, (Fig. 109,) and thus the suture, or line at which the edges of the germ unite to enclose the yolk, and which in the mammals forms the navel, is found in front. Another suture is found along the back, arising from the actual folding upwards of the upper surface of the germ, to form the dorsal cavity.

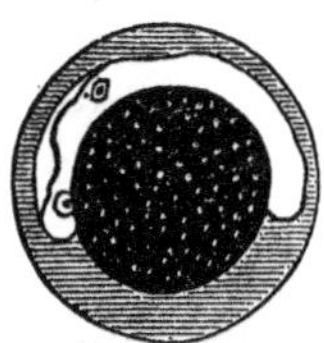

Fig. 109.

304. The embryo in the Articulata, on the contrary, lies with its back upon the yolk, as seen in the following figure, which represents an embryo of Podurella; consequently the yolk enters the body on that side; and the suture, which in the vertebrates is found on the belly, is here found on the back. In the Cephalopoda the yolk communicates with the lower side of the body, as in Vertebrates, but there is no dorsal cavity formed in them. In the other Mollusks, as also in the Worms, there is this peculiarity, that the whole yolk is changed at the beginning into the substance of the embryo; whilst in Vertebrates, and the higher Articulates and Mollusks, a part of it is reserved, till a later period, to be used for the nourishment of the embryo. Among Radiata, the germ is formed around the yolk, and seems to surround the whole of it, from the first.*

Fig. 110.

305. The development of the embryo of the vertebrated animals may be best observed in the eggs of fishes. Being

* These facts show plainly that the circumstance of embryos arising from the whole or a part of the yolk is of no systematic importance.

transparent, they do not require to be cut open, and, by sufficient caution, the whole series of embryonic changes may be observed upon the same individual, and thus the succession in which the organs appear be ascertained with precision; whereas, if we employ the eggs of birds, which are opaque, we are obliged to sacrifice an egg for each observation.

306. To illustrate these general views as to the development of the embryo, we will briefly describe the principal phases, as they have been observed in the White-fish of Europe, which belongs to the salmon family. The following magnified sections will illustrate this development, and show the period at which the different organs successively appear.

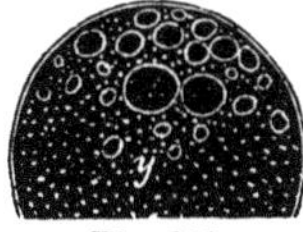

Fig. 111.

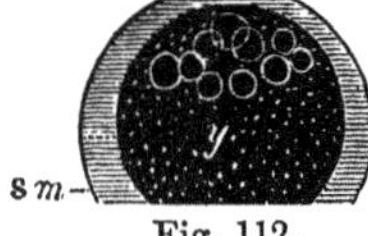

Fig. 112.

Fig. 113.

307. The egg, when laid, (Fig. 111,) is spherical, about the size of a small pea, and nearly transparent. It has no albumen, and the shell membrane is so closely attached to the membrane of the yolk, that they cannot be distinguished. Oil-like globules are scattered through the mass of the yolk, or grouped into a sort of disk, under which lies the germinative vesicle. The first change in such an egg occurs a few hours after it has been laid, when the shell membrane separates from the yolk membrane, in consequence of the absorption of a quantity of water, (Fig. 112,) by which the egg increases in size. Between the shell membrane (*s m*) and the yolk, (*y*,) there is now a considerable transparent space, which corresponds, in some respects, to the albumen found in the eggs of birds.

308. Soon afterwards we see, in the midst of the oil-like

globules, a swelling in the shape of a transparent vesicle, (Fig. 113, *g*,) composed of very delicate cells. This is the first indication of the *germ*. This swelling rapidly enlarges until it envelops a great part of the yolk, when a depression

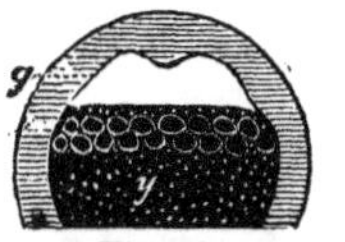

Fig. 114.

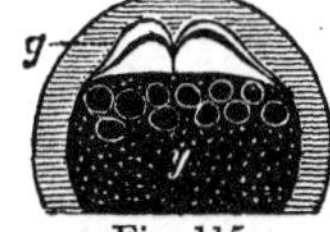

Fig. 115.

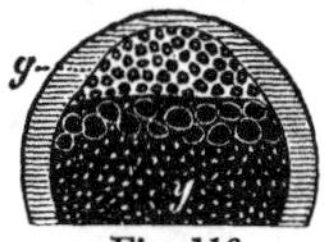

Fig. 116.

is formed upon it, (Fig. 114.) This depression becomes by degrees a deep furrow, and soon after a second furrow appears at right angles with the former, so that the germ now presents four elevations, (Fig. 115.) The subdivision goes on in this way, during the second and third days, until the germ is divided into numerous little spheres, giving the surface the appearance of a mulberry, (Fig. 116.) This appearance, however, does not long continue; at the end of the third day, the fissures again disappear, and leave no visible traces. After this, the germ continues to extend as an envelop around the yolk, which it at last entirely encloses.

309. On the tenth day, the first outlines of the embryo begin to appear, and we soon distinguish in it a depression between two little ridges, whose edges constantly approach

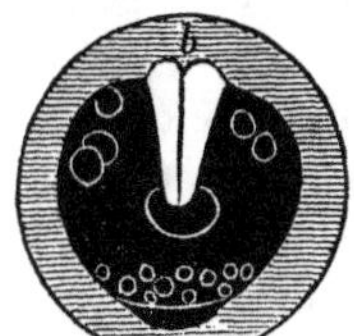

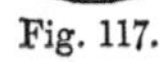
Fig. 117.

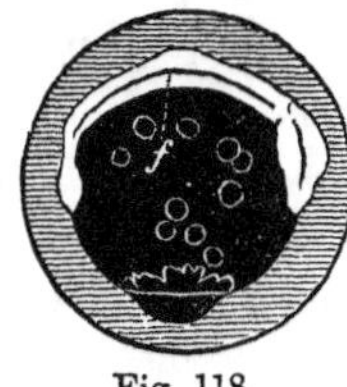

Fig. 118.

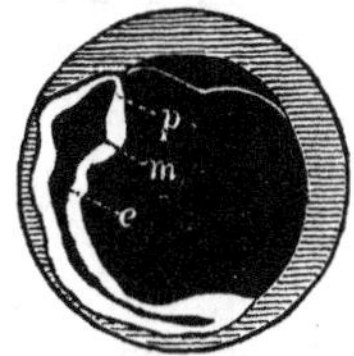

Fig. 119.

each other until they unite and form a canal, (Fig. 117, *b*,) as has been before shown, (Fig. 107.) At the same time, an enlargement at one end of the furrow is observed. This is the rudiment of the head, (Fig. 118,) in which may soon be distinguished traces of the three divisions of the brain, (Fig. 119,) corresponding to the senses of sight, (*m*,) hearing, (*e*,) and smell, (*p*.)

310. Towards the thirteenth day, we see a transparent, cartilaginous cord, in the place afterwards occupied by the back-bone, composed of large cells, on which transverse

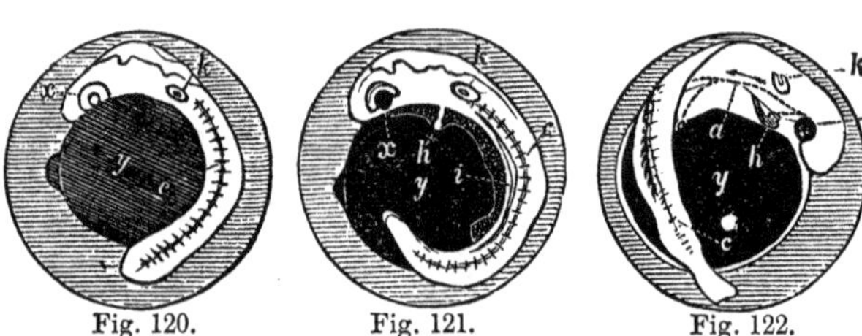

Fig. 120. Fig. 121. Fig. 122.

divisions are successively forming, (Figs. 120, 121, *c*.) This is the dorsal cord, a part of which, as we have before seen, is common to all embryos of vertebrated animals. It always precedes the formation of the back-bone; and in some fishes, as the sturgeon, this cartilaginous or embryonic state is permanent through life, and no true back-bone is ever formed. Soon after, the first rudiments of the eye appear, in the form of a fold in the external membrane of the germ, in which the crystalline lens (Fig. 121, *x*) is afterwards formed. At the same time we see, at the posterior part of the head, an elliptical vesicle, which is the rudiment of the ear. At this period, the distinction between the upper and the lower layer of the germ is best traced; all the changes mentioned above appertaining to the upper layer.

311. After the seventeenth day, the lower layer divides into two sheets, the inferior of which becomes the intestine.

The heart shows itself about the same time, under the form of a simple cavity, (Fig. 121, *h*,) in the midst of a mass of cells belonging to the middle or vascular layer. As soon as the cavity of the heart is closed in, regular motions of contraction and expansion are perceived, and the globules of blood are seen to rise and fall in conformity with these motions.

312. There is as yet, however, no circulation. It is not until the thirtieth day that its first traces are manifest in the existence of two currents, one running towards the head, the other towards the trunk, (Fig. 122,) with similar returning currents. At this time the liver begins to be formed. Meanwhile, the embryo gradually disengages itself, at both ends, from its adherence to the yolk; the tail becomes free, and the young animal moves it in violent jerks.

313. The embryo, although still enclosed in the egg, now unites all the essential conditions for the exercise of the functions of animal life. It has a brain, an intestine, a pulsating heart and circulating blood, and it moves its tail spontaneously. But the forms of the organs are not yet complete nor have they yet acquired the precise shape that characterizes the class, the family, the genus, and the species. The young White-fish is as yet only a vertebrate animal in general, and might as well be taken for the embryo of a frog.

314. Towards the close of the embryonic period, after the fortieth day, the embryo acquires a more definite shape. The head is more completely separated from the yolk, the jaws protrude, and the nostrils approach nearer and nearer to the end of the snout; divisions are formed in the fin which surrounds the body; the anterior limbs, which were indicated only by a small protuberance, assume the shape of fins; and finally, the openings of the gills appear, one after the other, so that we cannot now fail to recognize the type of fishes.

315. In this state, the young white-fish escapes from the

egg, about the sixtieth day after it is laid, (Fig. 123,) but its development is still incomplete. The outlines are yet too indistinct to indicate the genus and the species to which the fish belongs; at most we distinguish its order only. The opercula or gill-covers are not formed; the teeth are wanting; the fins have as yet no rays; the mouth is underneath, and it is some time before it assumes its final position at the most projecting point of the head. The remainder of the yolk is suspended from the belly, in the form of a large bladder, but it daily diminishes in size, until it is at length completely taken into the animal, (304.) The duration of these metamorphoses varies extremely in different fishes; some accomplish it in the course of a few days, while in others, months are required.

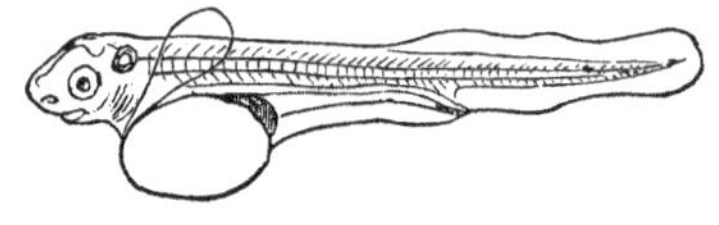

Fig. 123.

---

315 *a*. In frogs and all the naked reptiles, the development is very similar to that of fishes. It is somewhat different in the scaly reptiles, (snakes, lizards, and turtles,) which have peculiar membranes surrounding and protecting the embryo during its growth. From one of these envelopes, the allantoïs, (Fig. 125, *a*,) is derived their common name of *Allantoïdian Vertebrates*, in opposition to the naked reptiles and fishes, which are called *Anallantoïdian*.

315 *b*. The Allantoïdian Vertebrates differ from each other in several essential peculiarities. Among Birds, as well as in the scaly

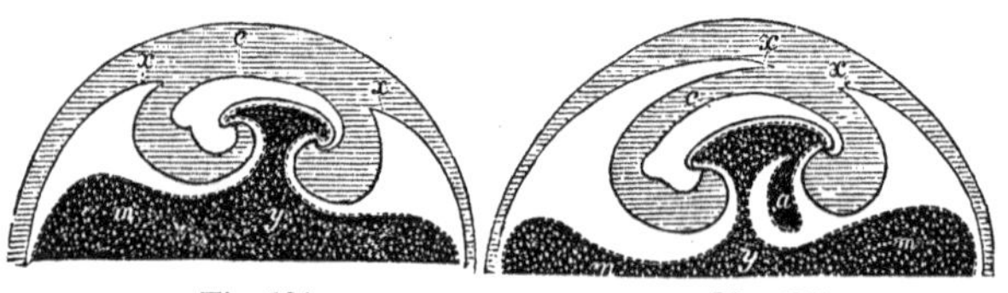

Fig. 124. Fig. 125.

reptiles, we find at a certain epoch, when the embryo is already dis-

316. As a general fact, it should be further stated, that the envelopes which protect the egg, and also the embryo, are the more numerous and complicated as animals belong to a higher class, and produce a smaller number of eggs. This is particularly evident when contrasting the innumerable eggs of fishes, discharged almost without protection

---

engaging itself from the yolk, a fold rising around the body from the upper layer of the germ, so as to present, in a longitudinal section, two prominent walls, (Fig. 124, *x x*.) These walls, converging from all sides upwards, rise gradually till they unite above the middle of the back, (Fig. 125.) When the junction is effected, which in the hen's egg takes place in the course of the fourth day, a cavity is formed between the back of the embryo (Fig. 126, *e*) and the new membrane, whose walls are called the *amnios*. This cavity becomes filled with a peculiar liquid, the *amniotic water*.

315 *c*. Soon after the embryo has been enclosed in the amnios, a shallow pouch forms from the mucous layer, below the posterior extremity of the embryo, between the tail and the vitelline mass. This pouch, at first a simple little sinus, (Fig. 125, *a*,) grows larger and larger, till it forms an extensive sac, the *allantois*, turning backwards and upwards, so as completely to separate the two plates of the amnios, (Fig. 126, *a*,) and finally enclosing the whole embryo, with its

Fig. 126.

amnios, in another large sac. The tubular part of this sac, which is nearest the embryo, is at last transformed into the urinary bladder. The heart (*h*) is already very large, with mniute arterial threads

into the water, with the well-protected eggs of birds, and still more with the growth of young mammals within the body of the mother.

317. But neither in fishes, nor in reptiles, nor in birds, does the vitelline membrane, or any other envelope of the egg, take any part in the growth of the embryo; while on the

---

passing off from it. At this period there exist true gills upon the sides of the neck, and a branchial respiration goes on.

315 *d*. The development of mammals exhibits the following peculiarities. The egg is exceedingly minute, almost microscopic, although composed of the same essential elements as those of the lower animals. The vitelline membrane, called *chorion*, in this class of animals, is comparatively thicker, (Fig. 127, *v*,) always soft, surrounded by peculiar cells, being a kind of albumen. The chorion soon grows proportionally larger than the vitelline sphere itself, (Fig. 128, *y*,) so as no longer to invest it directly, being separated from it by an empty space, (*k*.) The germ is formed in the same position as in the other classes of Vertebrates, namely, at the top of the vitellus, (Fig. 129;) and here also two layers may be distinguished, the upper or *serous* layer, (*s*,) and the lower or *mucous* layer, (*m*.) As it gradually enlarges, the surface of the chorion becomes covered with little fringes, which, at a later epoch, will be attached to the mother by means of similar fringes arising from the walls of the matrix, or organ which contains the embryo.

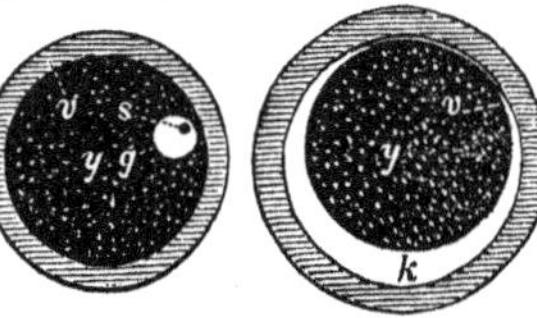

Fig. 127. Fig. 128.

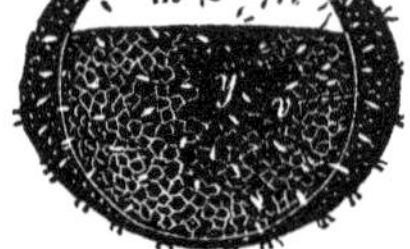

Fig. 129. Fig. 130.

315 *e*. The embryo itself undergoes, within the chorion changes

contrary, in the mammals, the chorion, which corresponds to the vitelline membrane, is vivified, and finally becomes attached to the maternal body, thus establishing a direct connection between the young and the mother; a connection which is again renewed in another mode, after birth, by the process of nursing.

---

similar to those described in birds: its body and its organs are formed in the same way; an amnios encloses it, and an allantoïs grows out of the lower extremity of the little animal. As soon as the allantoïs has surrounded the embryo, its blood vessels become more and more numerous, so as to extend into the fringes of the chorion, (Fig. 131, *p e*;) while, on the other hand, similar vessels from the mother extend into the corresponding fringes of the matrix, (*p m*,) but without directly communicating with those of the chorion. These two sorts of fringes soon become interwoven, so as to form an intricate organ filled with blood, called the ***placenta***, to which the embryo remains suspended until birth.

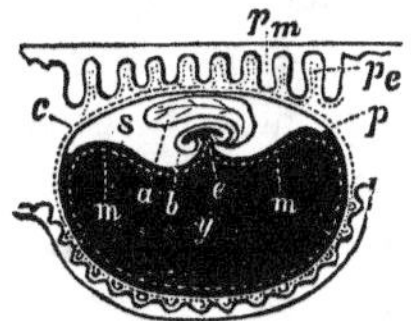

Fig. 131.

315 *f*. From the fact above stated, it is clear that there are three modifications of embryonic development among vertebrated animals, namely, that of fishes and naked reptiles, that of scaly reptiles and birds, and that of the mammals, which display a gradation of more and more complicated adaptation. In fishes and the naked reptiles, the germ simply encloses the yolk, and the embryo rises and grows from its upper part. In the scaly reptiles and birds there is, besides, an amnios arising from the peripheric part of the embryo and an allantoïs growing out of the lower cavity, both enclosing and protecting the germ.

SECTION III.

ZOÖLOGICAL IMPORTANCE OF EMBRYOLOGY.

318. As a general result of the observations which have been made, up to this time, on the embryology of the various classes of the Animal Kingdom, especially of the vertebrates, it may be said, that the organs of the body are successively formed in the order of their organic importance, the most essential being always the earliest to appear. In accordance with this law, the organs of vegetative life, the intestines and their appurtenances, make their appearance subsequently to those of animal life, such as the nervous system, the skeleton, &c.; and these, in turn, are preceded by the more general phenomena belonging to the animal as such.

319. Thus we have seen that, in the fish, the first changes relate to the segmentation of the yolk and the formation of the germ, which is a process common to all classes of animals. It is not until a subsequent period that we trace the dorsal furrow, which indicates that the forming animal will have a double cavity, and consequently belong to the division of the vertebrates; an indication afterwards fully confirmed by the successive appearance of the brain and the organs of sense. Later still, the intestine is formed, the limbs become evident, and the organs of respiration acquire their definite form, thus enabling us to distinguish with certainty the class to which the animal belongs. Finally, after the egg is hatched, the peculiarities of the teeth, and the shape of the extremities, mark the genus and species.

320. Hence the embryos of different animals resemble each other more strongly when examined in the earlier stages of their growth. We have already stated that, during

almost the whole period of embryonic life, the young fish and the young frog scarcely differ at all, (313:) so it is also with the young snake compared with the embryo bird. The embryo of the crab, again, is scarcely to be distinguished from that of the insect; and if we go still further back in the history of development, we come to a period when no appreciable difference whatever is to be discovered between the embryos of the various departments. The embryo of the snail, when the germ begins to show itself, is nearly the same as that of a fish or a crab. All that can be predicted at this period is, that the germ which is unfolding itself will become an animal; the class and the group are not yet indicated.

321. After this account of the history of the development of the egg, the importance of Embryology to the study of systematic Zoölogy cannot be questioned. For evidently, if the formation of the organs in the embryo takes place in an order corresponding to their importance, this succession must of itself furnish a criterion of their relative value in classification. Thus, those peculiarities that first appear should be considered of higher value than those that appear later. In this respect, the division of the Animal Kingdom into four types, the Vertebrates, the Articulates, the Mollusks, and the Radiates, corresponds perfectly with the gradations displayed by Embryology.

322. This classification, as has been already shown, (61,) is founded essentially on the organs of animal life, the nervous system and the parts belonging thereto, as found in the perfect animal. Now, it results from the above account, that in most animals the organs of animal life are precisely those that are earliest formed in the embryo; whereas those of vegetative life, on which is founded the division into classes, orders, and families, such as the heart, the respiratory apparatus, and the jaws, are not distinctly formed until afterwards. Therefore a classification, to be true and natural, must accord with the succession of organs in the embryonic development. This coincidence, while it corroborates the anatomical principles of Cuvier's classification of the Animal Kingdom, furnishes us with new proof that there is a general plan displayed in every kind of development.

323. Combining these two points of view, that of Embryology with that of Anatomy, the four divisions of the Animal Kingdom may be represented by the four figures which are to be found, at the centre of the diagram, at the beginning of the volume.

324. The type of Vertebrates, having two cavities, one above the other, the former destined to receive the nervous system, and the latter, which is of a larger size, for the intestines, is represented by a double crescent united at the centre, and closing above, as well as below.

325. The type of Articulata, having but one cavity, growing from below upwards, and the nervous system forming a series of ganglions, placed below the intestine, is represented by a single crescent, with the horns directed upwards.

326. The type of Mollusks having also but one cavity, the nervous system being a simple ring around the œsophagus, with ganglions above and below, from which threads go off to all parts, is represented by a single crescent with the horns turned downwards.

327. Finally, the type of Radiata, the radiating form of which is seen even in the youngest individuals, is represented by a star.

CHAPTER ELEVENTH.

PECULIAR MODES OF REPRODUCTION.

SECTION I.

GEMMIPAROUS AND FISSIPAROUS REPRODUCTION.

328. We have shown in the preceding chapter, that ovulation, and the development of embryos from eggs, is common to all classes of animals, and must be considered as the great process for the reproduction of species. Two other modes of propagation, applying, however, to only a limited number of animals, remain to be mentioned, namely, *gemmiparous* reproduction, or multiplication by means of buds, and *fissiparous* reproduction, or propagation by division; and also some still more extraordinary modifications yet involved in much obscurity.

329. *Reproduction by buds* occurs among the polyps, medusæ, and some of the infusoria. On the stalk, or even on the body of the Hydra, (Fig. 132,) and of many infusoria, there are formed buds, like those of plants. On close examination they are found to be young animals, at first very imperfectly formed, and communicating at the base with the parent body, from which they derive their nourishment. By degrees, the animal is developed; in most cases, the tube by which it is connected with the parent

Fig. 132.

withers away, and the animal is thus detached and becomes independent. Others remain through life united to the parent stalk, and, in this respect, present a more striking analogy to the buds of plants. But in the polyps, as in trees, budding is only an accessory mode of reproduction, which presupposes a trunk already existing, originally the product of ovulation.

330. *Reproduction by division*, or fissiparous reproduction, is still more extraordinary ; it takes place only in polyps and some infusoria. A cleft or fissure at some part of the body takes place, very slight at first, but constantly increasing in depth, so as to become a deep furrow, like that observed in the yolk, at the beginning of embryonic development ; at the same time the contained organs are divided and become double, and thus two individuals are formed of one, so similar to each other that it is impossible to say which is the parent and which the offspring. The division takes place sometimes vertically, as, for example, in Vorticella, (Fig. 133,) and in some Polyps, (Fig. 134,) and sometimes transversely. In some Infusoria, the Paramecia, for instance, this division occurs as often as three or four times in a day.

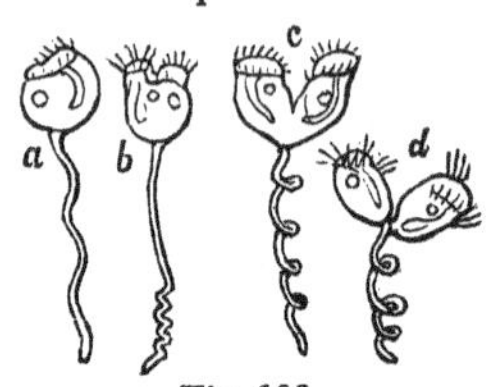

Fig. 133.

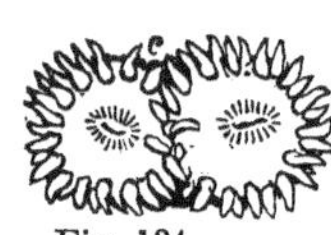

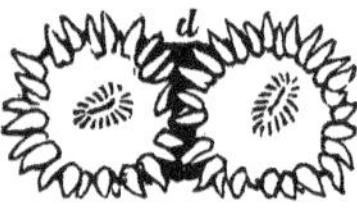

Fig. 134.

331. In consequence of this same faculty, many animals are able to reproduce various parts of their bodies when accidentally lost. It is well known that crabs and spiders, on losing a limb, acquire a new one. The same happens with the arms of the star-fishes. The tail of a lizard is also

readily reproduced. Salamanders even possess the faculty of reproducing parts of the head, including the eye with all its complicated structure. Something similar takes place in our own bodies, when a new skin is formed over a wound, or when a broken bone is reunited.

332. In some of the lower animals, this power of reparation is carried much farther, and applies to the whole body, so as closely to imitate fissiparous reproduction. If an earth-worm, or a fresh-water polyp, be divided into several pieces, the injury is soon repaired, each fragment speedily becoming a perfect animal. Something like this reparative faculty is seen in the vegetable kingdom, as well as the animal. A willow branch, planted in a moist soil, throws out roots below and branches above; and thus, after a time, assumes the shape of a perfect tree.

333. These various modes of reproduction do not exclude each other. All animals which propagate by gemmiparous or fissiparous reproduction also lay eggs. Thus the fresh-water polyps (Hydra) propagate both by eggs and by buds. In Vorticella, according to Ehrenberg, all three modes are found; it is propagated by eggs, by buds, and by division. Ovulation, however, is the most common mode of reproduction; the other modes, and also alternate reproduction, are only additional means employed by Nature to secure the perpetuation of the species.

SECTION II.

ALTERNATE AND EQUIVOCAL REPRODUCTION.

334. It is a matter of common observation, that individuals of the same species have the same general appearance, by which their peculiar organization is indicated. The transmission of these characteristics, from one generation to the next, is justly considered as one of the great laws of the Animal and Vegetable Kingdoms. It is, indeed, one of the points on which the definition of species is generally founded. We would, however, unhesitatingly adopt the new definition of Dr. S. G. Morton, who defines species to be "primordial organic forms."

335. But it does not follow that animals must resemble their parents in every condition, and at every epoch of their existence. On the contrary, as we have seen, this resemblance is very faint, in most species, at birth; and some, such as the caterpillar and the tadpole, undergo complete metamorphoses before attaining their final shape as the butterfly and frog. Nevertheless, we do not hesitate to refer *the tadpole and the frog to the same species;* and so with the caterpillar and the butterfly; because we know that there is the same individual observed in different stages of development.

336. There is, also, another series of cases, in which the offspring not only do not resemble the parent at birth, but, moreover, remain different during their whole life, so that their relationship is not apparent until a succeeding generation. The son does not resemble the father, but the grandfather; and in some cases the resemblance reappears only at the fourth or fifth generation, and even later. This singular mode of reproduction has received the name of *alternate generation.* The phenomena attending it have been of late the object of numerous scientific researches, which are the more deserving of our attention, as they furnish a solution to several problems alike interesting in a zoölogical and in a philosophical point of view.

337. Alternate generation was first observed among the Salpæ. These are marine mollusks, without shells, belonging to the family Tunicata. They are distinguished by the curious peculiarity of being united together in considerable numbers, so as to form long chains, which float in the sea,

the mouth, (*m*,) however, being free in each, (Fig. 135.)

Fig. 135.

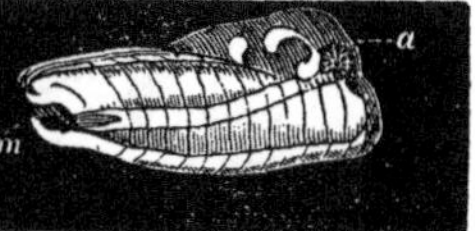

Fig. 136.

The individuals thus joined in floating colonies produce eggs; but in each animal there is generally but one egg formed, which is developed in the body of the parent, and from which is hatched a little mollusk, (Fig. 136,) which remains solitary, and differs in many respects from the parent. This little animal, on the other hand, does not produce eggs, but propagates by a kind of budding, which gives rise to chains already seen within the body of their parent, (*a*,) and these again bring forth solitary individuals, &c.

338. In some parasitic worms, alternate generation is accompanied by still more extraordinary phenomena, as is shown by the late discoveries of the Danish naturalist, Steenstrup. Among the numerous animals which inhabit stagnant pools, in which fresh-water shells, particularly Lymnea and Paludina, are found, there is a small worm, know to naturalists under the name of Cercaria, (Fig. 137.) When examined with a lens, it looks much like a tadpole, with a long tail, a triangular head, and a large sucker (*a*) in the middle of the body. Various viscera appear within, and, among others, a very distinct forked cord, (*c*,) which embraces the sucker, and which is thought to be the liver.

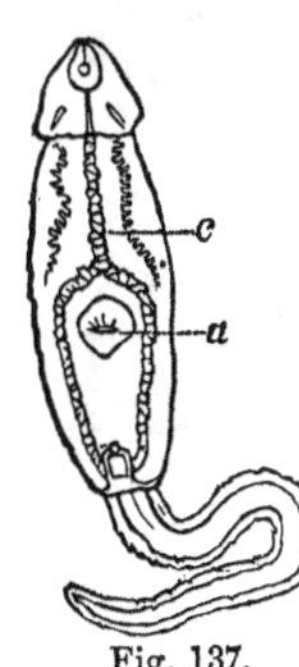

Fig. 137.

339. If we watch these worms, which always abound in company with the shells mentioned, we find them after a while attaching themselves, by means of their sucker, to the bodies of the mollusks. When

fixed, they soon undergo considerable alteration. The tail, which was previously employed for locomotion, is now useless, falls off, and the animal surrounds itself with a mucous substance, in which it remains nearly motionless, like the caterpillar on its transformation into the Pupa. If, however, after some time, we remove the little animal from its retreat, we find it to be no longer a Cercaria, but an intestinal worm, called Distoma, having the shape of Fig. **138**, with two suckers. The Distoma, therefore, is only a particular state of the Cercaria, or, rather, the Cercaria is only the larva of the Distoma.

Fig. 138.

340. What now is the origin of the Cercaria? The following are the results of the latest researches on this point. At certain periods of the year, we find in the viscera of the Limnea (one of the most common fresh-water mollusks) a quantity of little worms of an elongated form, with a well marked head, and two posterior projections like limbs, (Fig. **139**.) On examining these worms attentively, under the microscope, we discover that the cavity of their body is filled by a mass of other little worms, which a practised eye easily recognizes as young Cercariæ, the tail and the characteristic furcated organ (*a*) within it being distinctly visible, (Fig. 140.) These little embryos increase in size, distending the worm which contains them, and which seemingly has no other office than to protect and forward the development of the young Cercaria. It is, as it were, their living envelop. On this account, it has been called the *nurse*.

Fig. 139.

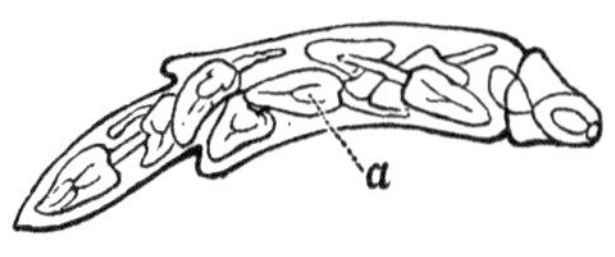

Fig. 140.

341. When they have reached a certain size, the young Cercariæ leave the body of the nurse, and move freely in the abdominal cavity of the mollusks, or escape from it into the water, to fix themselves, in their turn, to the body of another mollusk, and begin their transformations anew.

342. But this is not the end of the series. The nurses of the Cercaria are themselves the offspring of little worms of yet another kind. At certain seasons, we find in the viscera of the Limnea, worms somewhat like the nurses of Cercaria in shape, (Fig. 141,) but rather longer, more slender, and having a much more elongated stomach, (*s.*) These worms contain, in the hinder part of the body, little embryos, (*a*,) which are the young nurses, like Figures **139**, **140**. This generation has received the name of *grand-nurses*.

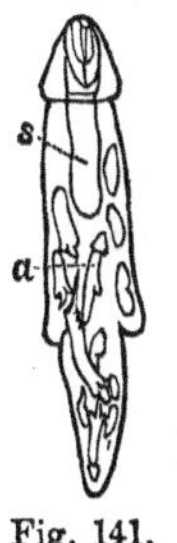

Fig. 141.

343. Supposing these grand-nurses to be the immediate offspring of the Distoma, (Fig. **138**,) as is probable, we have thus a quadruple series of generation. Four generations and one metamorphosis are required to evolve the perfect animal; in other words, the parent finds no resemblance to himself in any of his progeny, until he comes down to the great-grandson.

344. Among the Aphides, or plant-lice, the number of generations is still greater. The first generation, which is produced from eggs, soon undergoes metamorphoses, and then gives birth to a second generation, which is followed by a third, and so on; so that it is sometimes the eighth or ninth generation before the perfect animals appear as males and females, the sexes being then for the first time distinct, and the males provided with wings. The females lay eggs, which are hatched the following year, to repeat the same succession. Each generation is an additional step towards the perfect state; and, as each member of the succession is an incomplete animal, we cannot better explain their office, than by considering them analogous to the larvæ of the Cercaria, that is, as nurses.*

345. The development of the Medusæ is not less instructive. According to the observations of Sars, a Norwegian naturalist, the Medusa brings forth living young, which, after having burst the covering of the egg, swim about freely for some time in the body of the mother. When born, these animals have no resemblance whatever to the perfect Medusa. They are little cylindrical bodies, (Fig. 142, *a*,) much resembling infusoria, and, like them, covered with minute cilia, by means of which they swim with much activity.

346. After swimming about freely in the water for some days, the little animal fixes itself by one extremity, (Fig. 142, *e*.) At the opposite extremity a depression is gradu-

---

* There is a certain analogy between the larvæ of the plant-louse (Aphis) and the neuters or working ants and bees. This analogy has given rise to various speculations, and, among others, to the following theory, which is not without interest. The end and aim of all alternate generation, it is said, is to favor the development of the species in its progress towards the perfect state. Among the plant-lice, as among all the nurses, this end is accomplished by means of the body of the nurse. Now, a similar end is accomplished by the working ants and bees, only, instead of being performed as an organic function, it is turned into an outward activity, which makes them instinctively watch over the new generation, nurse and take care of it. It is no longer the body of the nurse, but its own instincts, which become the instrument of the development. This seems to receive confirmation from the fact that the working bees, like the plant-lice, are barren females. The attributes of their sex, in both, seem to consist only in their solicitude for the welfare of the new generation, of which they are the natural guardians, but not the parents. The task of bringing forth young is confided to other individuals, to the queen among the bees, and to the female of the last generation among the plant-lice. Thus the barrenness of the working bees, which seems an anomaly as long as we consider them complete animals, receives a very natural explanation so soon as we look upon them merely as nurses.

ally formed, the four corners (*b f*) become elongated, and, by degrees, are transformed into tentacles, (*c*.) These

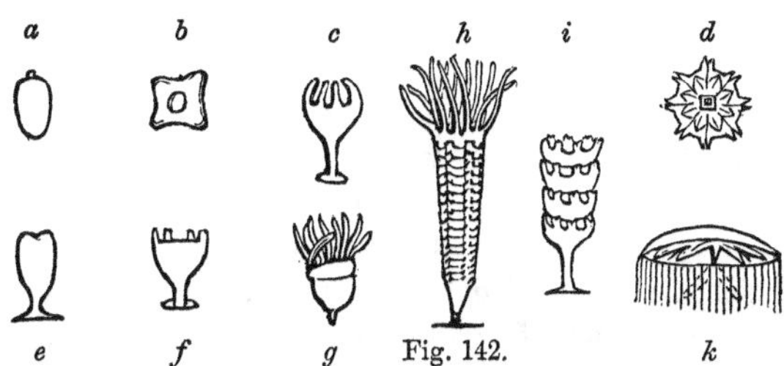

Fig. 142.

tentacles rapidly multiply, until the whole of the upper margin is covered with them, (*g*.) Then transverse wrinkles are seen on the body, at regular distances, appearing first above and extending downwards. These wrinkles, which are at first very slight, grow deeper and deeper, and, at the same time, the edge of each segment begins to be serrated, so that the animal presents the appearance of a pine cone, surmounted by a tuft of tentacles, (*h* ;) whence the name of Strobila, which was originally given to it, before it was known to be only a transient state of the jelly-fish. The separation constantly goes on, until at last the divisions are united by only a very slender axis, and resemble a pile of cups placed within each other, (*i*.) The divisions are now ready for separation; the upper ring first disengages itself, and then the others in succession.* Each segment (*d*) then continues its development by itself, until it becomes a complete Medusa, (*k* ;) while, according to recent researches, the basis or stalk remains and produces a new colony.

347. It is thus, by a series of metamorphoses, that the little animal which, on leaving the egg, has the form of the

* These free segments have been described as peculiar animals, under the name of *Ephyra*.

Infusoria, passes in succession through all the phases we have described. But the remarkable point in these metamorphoses is, that what was at first a single individual is thus transformed, by transverse division, into a number of entirely distinct animals, which is not the case in ordinary metamorphoses. Moreover, the upper segment does not follow the others in their development. Its office seems to be accomplished so soon as the other segments begin to be independent, being intended merely to favor their development, by securing and preparing the substances necessary to their growth. In this respect, it resembles the nurse of the Cercaria.

348. The Hydroid Polyps present phenomena no less numerous and strange. The Campanularia has a branching, plant-like form, with little cup-shaped cells on the ends and in the axils of the branches, each of which contains a little animal. These cups have not all the same organization. Those at the extremity of the branches, (*a*,) and which appear first, are furnished with long tentacles, wherewith they seize their food, (Fig. 143.) Those in the axils of the branches, and which appear late, are females, (*b*,) and have no such tentacles. Inside of the latter, little spherical bodies are found, each having several spots in the middle; these are the eggs. Finally, there is a third form, different from the two preceding, produced by budding from the female polyp, to which it in some sort belongs, (*c*.) It is within this that the eggs arrive, after having remained some time within the female. Their office seems to be to complete the incubation, for it is always within them that the eggs are hatched.

Fig. 143.

349. The little animal, on becoming free, has not the

slightest resemblance to the adult polyp. As in the young Medusa, the body is cylindrical, covered with delicate cilia. After having remained free for some time, the young animal fixes itself and assumes a flattened form. By degrees, a little swelling rises from the centre, which elongates, and at last forms a stalk. This stalk ramifies, and we soon recognize in it the animal of figure 143, with the three kinds of buds, which we may consider as three distinct forms of the same animal.

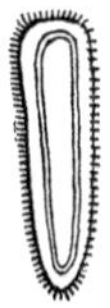

Fig. 144.

350. The development of Campanularia presents, in some respects, an analogy to what takes place in the reproduction of plants, and especially of trees. They should be considered as groups of individuals, and not as single individuals. The seed, which corresponds to the embryo of the Hydroid, puts forth a little stalk. This stalk soon ramifies by gemmiparous reproduction, that is, by throwing out buds which become branches. But ovulation, or reproduction by means of seeds, does not take place until an advanced period, and requires that the tree should have attained a considerable growth. It then produces flowers with pistils and stamens, that is, males and females, which are commonly united in one flower, but which in some instances are separated, as in the hickories, the elders, the willows, &c.*

* Several plants are endowed with organs similar to the third form of buds, as seen in the Campanularia; for example, the liverwort, (*Marchantia polymorpha*,) which has at the base of the cup a little receptacle, from the bottom of which little disk-like bodies are constantly forming, which, when detached, send out roots, and gradually become complete individuals. Besides that, we find in these animals, as in plants, the important peculiarity, that all the individuals are united in a common trunk, which is attached to the soil; and that all are intimately dependent on each other, as long as they remain united. And if we compare, in this point of view, the various species in which alternate reproduction has been observed, we find that the progress displayed in each type consists precisely in the increasing freedom of the individual in its various forms. At

SECTION III.

CONSEQUENCES OF ALTERNATE GENERATION.

351. These various examples of alternate generation render it evident, that this phenomenon ought not to be considered as an anomaly in Nature; but as the special plan of development, leading those animals in which it occurs to the highest degree of perfection of which they are susceptible. Moreover, it has been noticed among all types of invertebrated animals; while among the Vertebrates it is as yet unknown. It would seem that individual life in the lower animals is not defined within so precise limits as in the higher types; owing, perhaps, to the greater uniformity and independence of their constituent elements, the cells, and that, instead of passing at one stride as it were, through all the phases of their development, in order to accomplish it, they must either be born in a new form, as in the case of alternate generation, or undergo metamorphoses, which are a sort of second birth.

352. Many analogies may be discovered between alternate reproduction and metamorphosis. They are parallel lines that lead to the same end, namely, the development of the species. Nor is it rare to see them coëxisting in the same

first, we have all the generations united in a common trunk, as in the lower Polyps and in plants; then in the Medusæ and in some of the Hydroid Polyps the third generation begins to disengage itself. Among some of the intestinal worms, (the Distoma,) the third generation is enclosed within its nurse, and this, in its turn, is contained in the body of the grand-nurse, while the complete Distoma lives as a parasitic worm in the body of other animals, or even swims freely about in the larva state, as Cercaria. Finally, in the Plant-lice, all the generations, the nurses as well as the perfect animals, are separate individuals.

animal. Thus, in the Cercaria, we have seen an animal produced from a nurse afterwards transformed into a Distoma, by undergoing a regular metamorphosis.

353. In each new generation, as in each new metamorphosis, a real progress is made, and the form which results is more perfect than its predecessor. The nurse that produces the Cercaria is manifestly an inferior state, just as the chrysalis is inferior to the butterfly.

354. But there is this essential difference between the metamorphoses of the caterpillar and alternate reproduction, that, in the former case, the same individual passes through all the phases of development; whereas, in the latter, the individual disappears, and makes way for another, which carries out what its predecessors had begun. It would give a correct idea of this difference to suppose that the tadpole, instead of being itself transformed into a frog, should die, having first brought forth young frogs; or that the chrysalis should, in the same way, produce young butterflies. In either case, the young would still belong to the same species, but the cycle of development, instead of being accomplished in a single individual, would involve two or more acts of generation.

355. It follows, therefore, that the general practice of deriving the character of a species from the sexual forms alone, namely, the male and the female, is not applicable to all classes of animals; since there are large numbers whose various phases are represented by distinct individuals, endowed with peculiarities of their own. Thus, while in the stag the species is represented by two individuals only, stag and hind, the Medusa, on the other hand, is represented under the form of three different types of animals; the first is free, like the Infusoria, the second is fixed on a stalk, like a polyp, and the third again is free, consisting in its turn of male and female. In the Distoma, also, there are four

separate individuals, the grand-nurse, the nurse, the larva or Cercaria, and the Distoma, in which the sexes are not separate. Among the Aphides, the number is much greater still.

356. The study of alternate generation, besides making us better acquainted with the organization of the lower animals, greatly simplifies our nomenclature. Thus, in future, instead of enumerating the Distoma and the Cercaria, or the Strobila, the Ephyra, and the Medusa, as distinct animals, belonging to different classes and families, only the name first given to one of these forms will be retained, and the rest be struck from the pages of Zoölogy, as representing only the transitory phases of the same species.

357. Alternate generation always presupposes several modes of reproduction, of which the primary is invariably by ovulation. Thus, we have seen that the Polyps, the Medusa, the Salpa, &c., produce eggs, which are generally hatched within the mother. The subsequent generation, on the contrary, is produced in a different manner, as we have shown in the preceding paragraphs; as among the Medusæ, by transverse division; among the Polyps and Salpæ, by buds, &c.

358. The subsequent generations are, moreover, not to be regarded in the same light as those which first spring directly from eggs. In fact, they are rather phases of development, than generations properly so called; they are either without sex, or females whose sex is imperfectly developed. The nurses of the Distoma, the Medusa, and the Campanularia, are barren, and have none of the attributes of maternity, except that of watching over the development of the species, being themselves incapable of producing young.

359. Another important result follows from the above observations, namely, that the differences between animals which are produced by alternate generation are less, the

earlier the epoch at which we examine them. No two animals can be more unlike than an adult Medusa (Fig. 31) and an adult Campanularia, (Fig. 143;) they even seem to belong to different classes of the Animal Kingdom, the former being considered as an Acaleph, the latter as a Polyp. On the other hand, if we compare them when first hatched from the egg, they appear so much alike that it is with the greatest difficulty they can be distinguished. They are then little Infusoria, without any very distinct shape, and moving with the greatest freedom. The larvæ of certain intestinal worms, though they belong to a different department, have nearly the same form, at one period of their life. Farther still, this resemblance extends to plants. The spores of certain sea-weeds have nearly the same appearance as the young Polyp, or the young Medusa; and what is yet more remarkable, they are also furnished with cilia, and move about in a similar manner. But this is only a transient state. Like the young Campanularia and the young Medusa, the spore of the sea-weed is free for only a short time; soon it becomes fixed, and from that moment the resemblance ceases.

360. Are we to conclude, then, from this resemblance of the different types of animals at the outset of life, that there is no real difference between them; or that the two Kingdoms, the Animal and the Vegetable, actually blend, because their germs are similar? On the contrary, we think nothing is better calculated to strengthen the idea of the original separation of the various groups, as distinct and independent types, than the study of their different phases. In fact, a difference so wide as that between the adult Medusa and the adult Campanularia must have existed even in the young; only it does not show itself in a manner appreciable by our senses; the character by which they subsequently differ so much being not yet developed. To

deny the reality of natural groups, because of these early resemblances, would be to take the semblance for the reality. It would be the same as saying that the frog and the fish are one, because at one stage of embryonic life it is impossible, with the means at our command, to distinguish them.

361. The account we have above given of the development, the metamorphoses, and the alternate reproduction of the lower animals, is sufficient to undermine the old theory of *Spontaneous Generation*, which was proposed to account for the presence of worms in the bodies of animals, for the sudden appearance of myriads of animalcules in stagnant water, and under other circumstances rendering their occurrence mysterious. We need only to recollect how the Cercaria insinuates itself into the skin and the viscera of mollusks, (339, 342,) to understand how admission may be gained to the most inaccessible parts. Such beings occur even in the eye of many animals, especially of fishes; they are numerous in the eye of the common fresh-water perch of Europe. To the naked eye they seem like little white spots, (Fig. 145;) but when magnified, they have the form of Fig. 146.

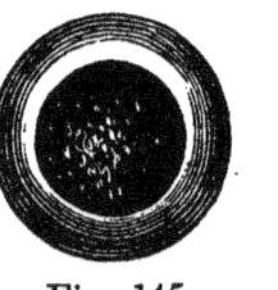
Fig. 145.

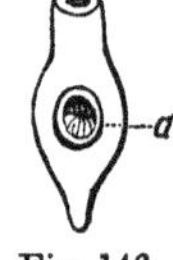

Fig. 146.

362. As to the larger intestinal worms found in other animals, the mystery of their origin has been entirely solved by recent researches. A single instance will illustrate their history. At certain periods of the year, the Sculpins of the Baltic are infested by a particular species of Tænia or tapeworm, from which they are free at other seasons. Mr. Eschricht found that, at certain seasons, the worms lose a great portion of the long chain of rings of which they are composed. On a careful examination, he found that each ring

contained several hundred eggs, which, on being freed from their envelop, float in the water. As these eggs are innumerable, it is not astonishing that the Sculpins should occasionally swallow some of them with their prey. The eggs, being thus introduced into the stomach of the fish, find conditions favorable to their development; and thus the species is propagated, and at the same time transmitted from one generation of the fish to another. The eggs which are not swallowed are probably lost.

363. All animals swallow, in the same manner, with their food, and in the water they drink, numerous eggs of such parasites, any one of which, finding in the intestine of the animal favorable conditions, may be hatched. It is probable that each animal affords the proper conditions for some particular species of worm; and thus we may explain how it is that most animals have parasites peculiar to themselves.

364. As respects the Infusoria, we also know that most of them, the Rotifera especially, lay eggs. These eggs, which are extremely minute, (some of them only $\frac{1}{12000}$ of an inch in diameter,) are scattered every where in great profusion, in water, in the air, in mist, and even in snow. Assiduous observers have not only seen the eggs laid, but moreover, have followed their development, and have seen the young animal forming in the egg, then escaping from it, increasing in size, and, in its turn, laying eggs. They have been able, in some instances, to follow them even to the fifth and sixth generation.

365. This being the case, it is much more natural to suppose that the Infusoria * are products of like germs, than

* In this connection, it ought to be remembered that a large proportion of the so-called Infusoria are not independent animals, but immature germs, belonging to different classes of the Animal Kingdom, and that many must be referred to the Vegetable Kingdom.

to assign to them a spontaneous origin altogether incompatible with what we know of organic development. Their rapid appearance is not at all astonishing, when we reflect that some mushrooms attain a considerable size in a few hours, but yet pass through all the phases of regular growth; and, indeed, since we have ascertained the different modes of generation among the lower animals, no substantial difficulties to the axiom "*omne vivum ex ovo*," (275,) any longer exist.

CHAPTER TWELFTH.

METAMORPHOSES OF ANIMALS.

366. UNDER the name of metamorphoses are included those changes which the body of an animal undergoes after its birth, and which are modifications, in various degrees, of its organization, form, and its mode of life. Such changes are not peculiar to certain classes, as has been so long supposed, but are common to all animals, without exception.

367. Vegetables also undergo metamorphoses, but with this essential difference, that in vegetables the process consists in an addition of new parts to the old ones. A succession of leaves, differing from those which preceded them, comes on each season; new branches and roots are added to the old stem, and woody layers to the trunk. In animals, the whole body is transformed, in such a manner that all the existing parts contribute to the formation of the modified body. The chrysalis becomes a butterfly; the frog, after having been herbivorous during its tadpole state, becomes carnivorous, and its stomach is adapted to this new mode of life; at the same time, instead of breathing by gills, it becomes an air-breathing animal; its tail and the gills disappear; lungs and legs are being developed, and, finally, it is to live and move on land.

368 The nature, the duration, and importance of metamorphoses, as also the epoch at which they take place, are infinitely varied. The most striking changes which naturally present themselves to the mind when we speak of me[illegible]or-

phoses, are those occurring in insects. Not merely is there a change of physiognomy and form observable, or an organ more or less formed, but their whole organization is modified. The animal enters into new relations with the external world, while, at the same time, new instincts are imparted to it. It has lived in water, and respired by gills; it is now furnished with air-tubes, and breathes in the atmosphere. It passes by, with indifference, objects which before were attractive, and its new instincts prompt it to seek conditions which would have been most pernicious during its former period of life. All these changes are brought about without destroying the individuality of the animal. The mosquito, which to-day haunts us with its shrill trumpet, and pierces us for our blood, is the same animal that, a few days ago, lived obscure and unregarded in stagnant water, under the guise of a little worm.

369. Every one is familiar with the metamorphoses of the silk-worm. On escaping from the egg, the little worm or caterpillar grows with great rapidity for twenty days, when it ceases to feed, spins its silken cocoon, casts its skin, and remains enclosed in its chrysalis state.* During this period of its existence, most extraordinary changes take place. The jaws with which it masticated mulberry leaves are transformed into a coiled tongue; the spinning organs are reduced; the gullet is lengthened and more slender; the stomach, which was nearly as long as the body, is now contracted into a short bag; the intestine, on the contrary, becomes elongated and narrow. The dorsal vessel is shortened. The ganglions of the thoracic region approach each other, and unite into a single mass. Antennæ and palpi are developed on the head, and instead of simple eyes appear compound ones.

* In the raising of silk-worms this period is not waited for, but the animal is killed as soon as it has spun its cocoon.

The muscles, which before were uniformly distributed, (159,) are now gathered into masses. The limbs are elongated, and wings spring forth from the thorax. More active motions then reappear in the digestive organs, and the animal, bursting the envelop of its chrysalis, issues in the form of a winged moth.

370 The different external forms which an insect may assume is well illustrated by one which is unfortunately too well known in this country, namely, the canker-worm. Its eggs are laid on posts and fences, or upon the branches of our apple-trees, elms, and other trees. They are hatched about the time the tender leaves of these trees begin to unfold.

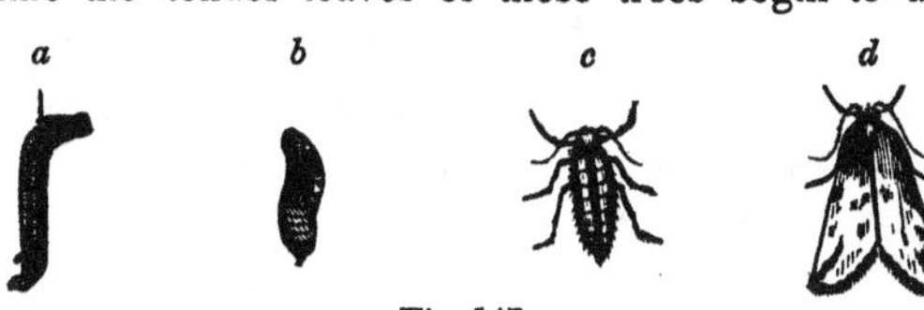

Fig. 147.

The caterpillar (*a*) feeds on the leaves, and attains its full growth at the end of about four weeks, being then not quite an inch in length. It then descends to the ground, and enters the earth to the depth of four or five inches, and having excavated a sort of cell, is soon changed into a chrysalis or nymph, (*b.*) At the usual time in the spring, it bursts the skin, and appears in its perfect state, under the form of a moth, (*d.*) In this species, however, only the male has wings. The perfect insects soon pair, the female (*c*) crawls up a tree, and, having deposited her eggs, dies.

371. Transformations no less remarkable are observed among the Crustacea. The metamorphoses in the family of Cirrhipedes are especially striking. It is now known that the barnacles, (Balanus,) which have been arranged among the mollusks, are truly crustaceans; and this result of modern researches has been deduced in the clearest manner from the

study of their transformations. The following figures represent the different phases of the duck-barnacle, (Anatifa.)

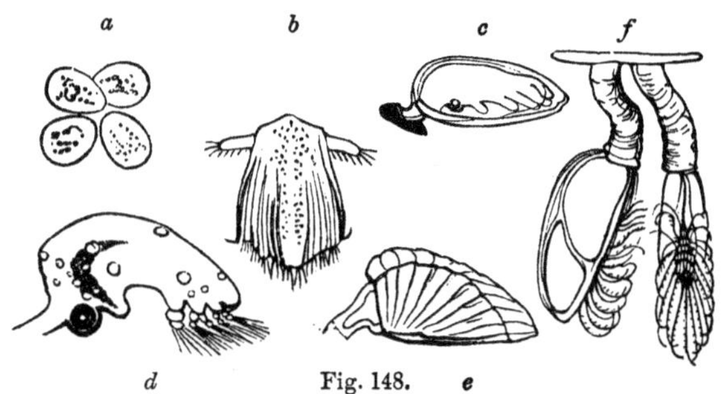

Fig. 148.

372. The Anatifa, like all crustacea, is reproduced by eggs, specimens of which, magnified ninety diameters, are represented in figure 148, *a*. From these eggs little animals issue, which have not the slightest resemblance to the parent. They have an elongated form, (*b*,) a pair of tenacles, and four legs, with which they swim freely in the water.

373. Their freedom, however, is of but short duration. The little animal soon attaches itself by means of its tentacles, having previously become covered with a transparent shell, through which the outlines of the body, and also a very distinct eye, are easily distinguished, (Fig. 148, *c*.) Figure 148, *d*, shows the animal taken out of its shell. It is plainly seen that the anterior portion has become considerably enlarged. Subsequently, the shell becomes completed, and the animal casts its skin, losing with it both its eyes and its tentacles. On the other hand, a thick membrane lines the interior of the shell, which pushes out and forms a stem, (*e*,) by means of which the animal fixes itself to immersed bodies, after the loss of its tentacles. This stem gradually enlarges, and the animal soon acquires a definite shape, such

as it is represented in figure 148, *f*, attached to a piece of floating wood.

374. There is, consequently, not only a change of organization in the course of the metamorphoses, but also a change of faculties and mode of life. The animal, at first free, becomes fixed; and its adhesion is effected by totally different organs at different periods of life, first by means of tentacles, which were temporary organs, and afterwards by means of a fleshy stem developed especially for that purpose.

375. The Radiata also furnish us with examples of various metamorphoses, especially among the star-fishes. A small species living on the coast of New England (*Echinaster sanguinolentus*) undergoes the following phases, (Fig. 149.)

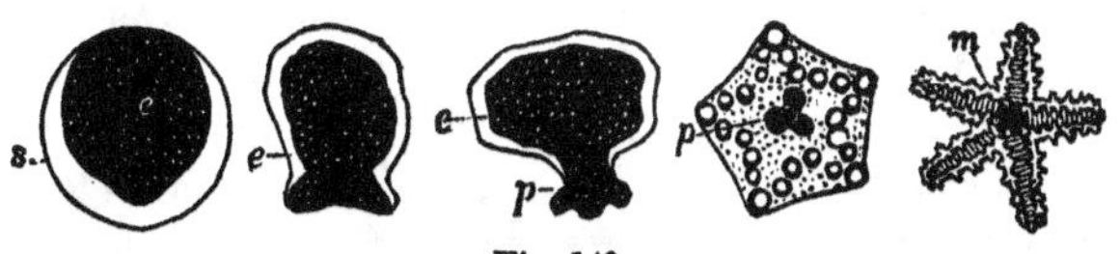

Fig. 149.

376. If the eggs are examined by the microscope, each one is found to contain a small, pear-shaped body, which is the embryo, (*e*,) surrounded by a transparent envelop. On escaping from the egg, the little animal has an oblong form, with a constriction at the base. This constriction becoming deeper and deeper forms a pedicle, (*p*,) which soon divides into three lobes. The disk also assumes a pentagonal form, with five double series of vesicles. The first rudiments of the rays are seen to form in the interior of the pentagon. At the same time, the peduncle contracts still more, being at last entirely absorbed into the cavity of the body, and the animal soon acquires its final form, (*m*.)

377. Analogous transformations take place in the Comatula. In early life (Fig. 150) it is fixed to the ground by a stem, but becomes detached at a certain epoch, and then floats freely in the sea, (Fig. 151.) On the other hand, the Polypi seem to follow a reverse course, many of them becoming fixed to the ground after having been previously free.

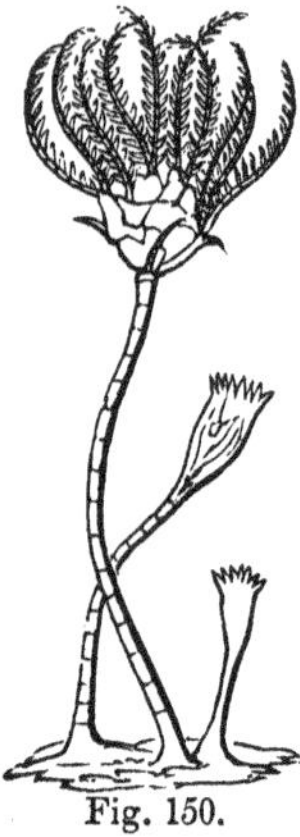
Fig. 150.

Fig. 151.

378. The metamorphoses of mollusks, though less striking, are not less worthy of notice. Thus, the oyster, with which we are familiar in its adhering shell, is free when young, like the clam (Mya) and most other shell-fishes. Others, which are at first attached or suspended to the gills of the mother, afterwards become free, as the Unio. Some naked Gasteropods, the Acteon or the Eolis, for example, are born with a shell, which they part with shortly after leaving the egg.

379. The study of metamorphoses is, therefore, of the utmost importance for understanding the real affinities of animals very different in appearance, as is readily shown by the following instances. The butterfly and the earth-worm seem, at the first glance, to have no relation whatever. They differ in their organization, no less than in their outward appearance. But, on comparing the caterpillar and the worm, these two animals closely resemble each other. The analogy, however, is only transient; it lasts only during the larva state of the caterpillar, and is effaced as it

passes to the chrysalis and butterfly states. The latter becomes a more and more perfect animal, whilst the worm remains in its inferior state.

380. Similar instances are furnished by animals belonging to all the types of the Animal Kingdom. Who would think, at first glance, that a Barnacle or an Anatifa were more nearly allied to the crab than to the oyster? And, nevertheless, we have seen, (372,) in tracing back the Anatifa to its early stages, that it then bears a near resemblance to a little Crustacean, (Fig. 148, *d.*) It is only when full grown that it assumes its peculiar mollusk-like covering.

381. Among the Cuttle-fishes there are several, the Loligo, (Fig. 47,) for example, which are characterized by the form of their tentacles, the two interior being much longer than the others, and of a different form; whilst in others, as the Octopus, they are all equal. But if we compare the young, we find that in both animals the tentacles are all equal, though they differ in number. The inequality in the tentacles is the result of a further development.

382. Among the Radiata, the Pentacrinus and the Comatula exemplify the same point. The two are very different when full grown, the latter being a free-swimming star-fish, (Fig. 151,) while the former is attached to the soil, like a Polyp. But we have seen (377) that the same is the case with Comatula in its early period; and that, in consequence of a further metamorphosis, it becomes disengaged from its stem, and floats freely in the water.

383. In the type of Vertebrates, the considerations drawn from metamorphoses acquire still greater importance in reference to classification. The Sturgeon and the White-fish, before mentioned, (306,) are two very different fishes; yet, taking into consideration their external form and bearing merely, it might be questioned which of the two should take the highest rank; whereas the doubt is very easily

resolved by an examination of their anatomical structure. The White-fish has a skeleton, and, moreover, a vertebral column, composed of firm bone. The Sturgeon, (Fig. 152,)

Fig. 152.

on the contrary, has no bone in the vertebral column, except the spines or apophyses of the vertebræ. The middle part, or body of the vertebra, is cartilaginous ; the mouth is transverse, and underneath the head ; and the caudal fin is unequally forked, while in the White-fish it is equally forked.

384. If, however, we observe the young White-fish just after it has issued from the egg, (Fig. 123,) the contrast will be less striking. At this period the vertebræ are cartilaginous, like those of the Sturgeon ; its mouth, also, is transverse and inferior, and its tail undivided ; at that period the White-fish and the Sturgeon are, therefore, much more alike. But this similarity is only transient ; as the White-fish grows, its vertebræ become ossified, and its resemblance to the Sturgeon is comparatively slight. As the Sturgeon has no such transformation of the vertebræ, and is, in some sense, arrested in its development, while the White-fish undergoes subsequent transformation, we conclude that, compared with the White-fish, it is really inferior in rank.

385. This relative inferiority and superiority strikes us still more when we compare with our most perfect fishes (the Salmon, the Cod) some of those worm-like animals, so different from ordinary fishes that they were formerly placed among the worms. The Amphioxus, represented of its natural size, (Fig. 153,) not only

Fig. 153.

has no bony skeleton, but not even a head, properly speak ing. Yet the fact that it possesses a dorsal cord, extending from one extremity of the body to the other, proves that it belongs to the type of Vertebrates. But as this peculiar structure is found only at a very early period of embryonic development, in other fishes, we conclude that the Amphioxus holds the very lowest rank in this class.

386. Nevertheless, the metamorphoses of animals after birth, will, in many instances, present but trifling modifications of the relative rank of animals, compared with those which may be derived from the study of changes previous to that period, as there are many animals which undergo no changes of great importance after their escape from the egg, and occupy, nevertheless, a high rank in the Zoölogical series, as, for example, Birds and Mammals. The question is, whether such animals are developed according to different plans, or whether their peculiarity in that respect is merely apparent. To answer this question, let us go back to the period anterior to birth, and see if some parallel may not be made out between the embryonic changes of these animals and the metamorphoses which take place subsequently to birth in others.

387. We have already shown that embryonic development consists in a series of transformations; the young animal enclosed in the egg differing at each period of its development, from what it was before. But because these transformations precede birth, and are, therefore, not generally observed, they are not less important. To be satisfied that these transformations are in every respect similar to those which follow birth, we have only to compare the changes which immediately precede birth with those which immediately follow it, and we shall readily perceive that the latter are simply a continuation of the former, till all are completed.

388. Let us recur to the development of fishes for illus-

tration. The young White-fish, as we have seen, (315,) is far from having acquired its complete development when born. The vertical fins are not yet separate; the mouth has not yet its proper position; the yolk has not yet retreated within the cavity of the body, but hangs below the chest in the form of a large bag. Much, therefore, remains to be changed before its development is complete. But the fact that it has been born does not prevent its future evolution, which goes on without interruption.

389. Similar inferences may be drawn from the development of the chicken. The only difference is, that the young chicken is born in a more mature state, the most important transformations having taken place during the embryonic period, while those to be undergone after birth are less considerable, though they complete the process begun in the embryo. Thus we see it, shortly after birth, completely changing its covering, and clothed with feathers instead of down; still later its crest appears, and its spurs begin to be developed.

390. In certain Mammals, known under the name of Marsupials, (the Opossum and Kangaroo,) the link between the transformations which take place before birth, and those that occur at a later period, is especially remarkable. These animals are brought into the world so weak and undeveloped that they have to undergo a second gestation, in a pouch with which the mother is furnished, and in which the young remain, each one fixed to a teat, until they are entirely developed. Even those animals which are born nearest to the complete state, undergo, nevertheless, embryonic transformations. Ruminants acquire their horns; and the lion his mane. Most mammals, at birth, are destitute of teeth, and incapable of using their limbs; and all are dependent on the mother and the milk secreted by her, until the stomach is capable of digesting other aliments.

391. If it be thus shown that the transformations which take place in the embryo are of the same nature, and of the same importance, as those which occur afterwards, the circumstance that some precede and others succeed birth cannot mark any radical distinction between them. Both are processes of the life of the individual. Now, as life does not commence at birth, but goes still farther back, it is quite clear that the modifications which supervene during the former period are essentially the same as, and continuous with, the later ones; and hence, that metamorphoses, far from being exceptional in the case of Insects, are one of the general features of the Animal Kingdom.

392. We are, therefore, perfectly entitled to say that all animals, without exception, undergo metamorphoses. Were it not so, we should be at a loss to conceive why animals of the same division present such wide differences; and that there should be, as in the class of Reptiles, some families that undergo important metamorphoses, (the frogs, for example,) and others in which nothing of the kind is observed after birth, (the Lizards and Tortoises.)

393. It is only by connecting the two kinds of transformations, namely, those which take place before, and those after birth, that we are furnished with the means of ascertaining the relative perfection of an animal; in other words, these transformations become, under such circumstances, a natural key to the gradation of types. At the same time, they will force upon us the conviction that there is an immutable principle presiding over all these changes, and regulating them in a peculiar manner in each animal.

394. These considerations are exceedingly important, not only from their bearing upon classification, but not less so from the application which may be made of them to the study of fossils. If we examine attentively the fishes that have been found in the different strata of the earth, we remark that

those of the most ancient deposits have, in general, preserved only the apophyses of their vertebræ, whilst the vertebræ themselves are wanting. Were the Sturgeons of the American rivers to become petrified, they would be found in a similar state of preservation. As the apophyses are the only bony portions of the vertebral column, they alone would be preserved. Indeed, fossil Sturgeons are known, which are in precisely this condition.

395. From the fact above stated, we may conclude that the oldest fossil fishes did not pass through all the metamorphoses which our osseous fishes undergo ; and, consequently, that they were inferior to analogous species of the present epoch which have bony vertebræ. Similar considerations apply to the fossil crustacea and to the fossil Echinoderms, when compared with living ones, and will, probably, be true of all classes of the Animal Kingdom, when fully studied as to their geological succession.

CHAPTER THIRTEENTH.

GEOGRAPHICAL DISTRIBUTION OF ANIMALS.

SECTION I.

GENERAL LAWS OF DISTRIBUTION.

396. No animal, excepting man, inhabits every part of the surface of the earth. Each great geographical or climatal region is occupied by some species not found elsewhere; and each animal dwells within certain limits, beyond which it does not range while left to its natural freedom, and within which it always inclines to return, when removed by accident or design. Man alone is a cosmopolite. His domain is the whole earth. For him, and with a view to him, it was created. His right to it is based upon his organization and his relation to Nature, and is maintained by his intelligence and the perfectibility of his social condition.

397. A group of animals which inhabits any particular region, embracing all the species, both aquatic and terrestrial, is called its FAUNA; in the same manner as the plants of a country are called its Flora. To be entitled to this name, it is not necessary that none of the animals composing the group should be found in any other region; it is sufficient that there should be peculiarities in the distribution of the families, genera, and species, and in the preponderance of certain types over others, sufficiently prominent to impress upon a region well-marked features. Thus, for example, in the islands of the Pacific are found terrestrial animals, altogether peculiar, and not found on the nearest continents. There are numerous animals in New Holland differing from any found on the continent of Asia, or, indeed, on any other part of the earth. If, however, some species inhabiting both shores of a sea which separates two terrestrial regions are found to be alike, we are not to conclude that those regions have the same Fauna, any more than that the Flora of Lapland and England are alike, because some of the sea-weeds found on both their shores are the same.

398. There is an evident relation between the fauna of any locality and its temperature, although, as we shall hereafter see, similar climates are not always inhabited by similar animals, (401, 402.) Hence the faunas of the two hemispheres have been distributed into three principal divisions, namely, the arctic, the temperate, and the tropical faunas; in the same manner as we have arctic, temperate, and tropical floras. Hence, also, animals dwelling at high elevations upon mountains, where the temperature is much reduced, resemble the animals of colder latitudes, rather than those of the surrounding plains.

399. In some respects, the peculiarities of the fauna of a region depend upon its flora, at least so far as land animals are concerned; for herbivorous animals will exist only where there is an adequate supply of vegetable food. But taking the terrestrial and aquatic animals together, the limitation of a fauna is less intimately dependent on climate than that of a flora. Plants, in truth, are for the most part terrestrial, (marine plants being relatively very few,) while animals are chiefly aquatic. The ocean is the true home of the Animal Kingdom; and while plants, with the exception of the lichens and mosses, become dwarfed, or perish under the influence of severe cold, the sea teems with animals of all classes, far beyond the extreme limit of flowering plants.

400. The influence of climate, in the colder regions, acts merely to induce a greater uniformity in the species of animals. Thus the same animals inhabit the northern polar regions of the three continents. The polar bear is the same in Europe, Asia, and America, and so are also a great many birds. In the temperate regions, on the contrary, the species differ on each of the continents, but they still preserve the same general features. The types are the same, but they are represented by quite different species. In consequence of these general resemblances, the first colonists of New England erroneously applied the names of European species to American animals. Similar differences are observed in distant regions of the same continent, within the same parallels of latitude. The animals of Oregon and of California are not the same as those of New England. The difference, in certain respects, is even greater than between the animals of New England and Europe. In like manner, the animals of temperate Asia differ more from those of Europe than they do from those of America.

401. Under the torrid zone, the Animal Kingdom, as well as the Vegetable, attains its highest development. The animals of the tropics are not only different from those of the temperate zone, but, moreover, they present the greatest variety among themselves. The most gracefully proportioned forms are found by the side of the most grotesque, decked with every combination of brilliant coloring. At the same time, the contrast between the animals of different continents is more marked; and, in many respects, the animals of the different tropical faunas differ not less from each other than from those of the temperate or frozen zones. Thus, the fauna of Brazil varies as much from that of Central Africa as from that of the United States.

402. This diversity upon different continents cannot depend simply on any influence of the climate of the tropics

if it were so uniformity ought to be restored in proportion as we recede from the tropics towards the antarctic temperate regions. But, instead of this, the differences continue to increase ; — so much so, that no faunas are more in contrast than those of Cape Horn, the Cape of Good Hope, and New Holland. Hence, other influences must be in operation besides those of climate ; — influences of a higher order, which are involved in a general plan, and intimately associated with the development of life on the surface of the earth.

403. Faunas are more or less distinctly limited, according to the natural features of the earth's surface. Sometimes two faunas are separated by an extensive chain of mountains, like the Rocky Mountains. Again, a desert may intervene, like the desert of Sahara, which separates the fauna of Central Africa from that of the Atlas and the Moorish coast, the latter being merely an appendage to the fauna of Europe. But the sea effects the most complete limitation. The depths of the ocean are quite as impassable for marine species as high mountains are for terrestrial animals. It would be quite as difficult for a fish or a mollusk to cross from the coast of Europe to the coast of America, as it would be for a reindeer to pass from the arctic to the antarctic regions, across the torrid zone. Experiments of dredging in very deep water have also taught us that the abyss of the ocean is nearly a desert. Not only are no materials found there for sustenance, but it is doubtful if animals could sustain the pressure of so great a column of water, although many of them are provided with a system of pores, (260,) which enables them to sustain a much greater pressure than terrestrial animals.

404. When there is no great natural limit, the transition from one fauna to another is made insensibly. Thus, in passing from he arctic to the temperate regions of North

America, one species takes the place of another, a third succeeds the second, and so on, until finally the fauna is found to be completely changed, though it is not always possible to mark the precise line which divides the one from the other.

405. The range of species does not at all depend upon their powers of locomotion; if it were so, animals which move slowly and with difficulty would have a narrow range, whilst those which are very active would be widely diffused. Precisely the reverse of this is actually the case. The common oyster extends at least from the St. Lawrence to the Carolinas; its range is consequently very great; much more so than that of some of the fleet animals, as, for instance, the Moose. It is even probable that the very inability of the oyster to travel really contributes to its diffusion, inasmuch as, having once spread over extensive grounds, there is no chance of its return to a former limitation, inasmuch as, being fixed, and consequently unable to choose positions for its eggs, they must be left to the mercy of currents; while Fishes, by depositing their eggs in the bays and inlets of the shore, undisturbed by currents and winds, secure them from too wide a dispersion.

406. The nature of their food has an important bearing upon the grouping of animals, and upon the extent of their distribution. Carnivorous animals are generally less confined in their range than herbivorous ones; because their food is almost every where to be found. The herbivora, on the contrary, are restricted to the more limited regions corresponding to the different zones of vegetation. The same remark may be made with respect to Birds. Birds of prey, such as the eagle and vulture, have a much wider range than the granivorous and gallinaceous birds. Still, notwithstanding the facilities they have for change of place, even the birds that wander widest recognize limits which they do

not overstep. The Condor of the Cordilleras does not de scend into the temperate regions of the United States; and yet it is not that he fears the cold, since he is frequently known to ascend even above the highest summits of the Andes, and disappears from view where the cold is most intense. Nor can it be from lack of prey.

407. Again, the peculiar configuration of a country sometimes determines a peculiar grouping of animals, into what may be called local faunas. Such, for example, are the prairies of the West, the Pampas of South America, the Steppes of Asia, the Deserts of Africa; — and, for marine animals, the basin of the Caspian. In all these localities, animals are met with which exist only there, and are not found except under those particular conditions.

408. Finally, to obtain a true picture of the zoölogical distribution of animals, not the terrestrial types alone, but the marine species, must also be included. Notwithstanding the uniform nature of the watery element, the animals which dwell in it are not dispersed at random; and though the limits of the marine may be less easily defined than those of the terrestrial faunas, still, marked differences between the animals of great basins are not less observable. Properly to apprehend how marine animals may be distributed into local faunas, it must be remembered that their residence is not in the high sea, but along the coasts of continents and on soundings. It is on the Banks of Newfoundland, and not in the deep sea, that the great cod-fishery is carried on; and it is well known that when fishes migrate, they run along the shores. The range of marine species being, therefore, confined to the vicinity of the shores, their distribution must be subjected to laws similar to those which regulate the terrestrial faunas. As to the fresh-water fishes, not only do the species vary in the different zones, but even the different rivers of the same region have species peculiar to them, and

not found in neighboring streams. The garpikes (Lepidosteus) of the American rivers afford a striking example of this kind.

409. A very influential cause in the distribution of aquatic animals is the depth of the water; so that several zoölogical zones, receding from the shore, may be defined, according to the depth of water; much in the same manner as we mark different zones at different elevations in ascending mountains, (398.) The Mollusks, and even the Fishes found near the shore in shallow water, differ, in general, from those living at the depth of twenty or thirty feet, and these again are found to be different from those which are met with at a greater depth. Their coloring, in particular, varies, according to the quantity of light they receive, as has also been shown to be the case with the marine plants.

410. It is sometimes the case that one or more animals are found upon a certain chain of mountains, and not elsewhere; as, for instance, the Mountain Sheep (*Ovis montana*) upon the Rocky Mountains, or the Chamois and the Ibex upon the Alps. The same is also the case on some of the wide plains or prairies. This, however, does not entitle such regions to be considered as having an independent fauna, any more than a lake is to be regarded as having a peculiar fauna, exclusive of the animals of the surrounding country, merely because some of the species found in the lake may not ascend the rivers emptying into it. It is only when the whole group of animals inhabiting such a region has such peculiarities as to give it a distinct character, when contrasted with animals found in surrounding regions, that it is to be regarded as a separate fauna. Such, for example, is the fauna of the great steppe, or plain of Gobi, in Asia; and such indeed that of the chain of the Rocky Mountains may prove to be, when the animals inhabiting them shall be better known.

411. The migration of animals might at first seem to present a serious difficulty in determining the character or the limits of a fauna; but this difficulty ceases, if we regard the country of an animal to be the place where it makes its habitual abode. As to Birds, which of all animals wander farthest, it may be laid down as a rule, that they belong to the zone in which they breed. Thus, the gulls, many of the ducks, mergansers, and divers, belong to the boreal regions, though they pass a portion of the year with us. On the other hand, the swallows and martins, and many of the gallinaceous birds belong to the temperate faunas, notwithstanding their migration during winter to the confines of the torrid zone. This rule does not apply to the fishes who annually leave their proper home, and migrate to a distant region merely for the purpose of spawning. The Salmon, for example, comes down from the North, to spawn on the coast of Maine and Nova Scotia.

412. Few of the Mammals, and these mostly of the tribe of Rodents, make extensive migrations. Among the most remarkable of these are the Kamtschatka rats. In Spring they direct their course westward, in immense troops; and, after a very long journey, return again in Autumn to their quarters, where their approach is anxiously awaited by the hunters, on account of the fine furs to be obtained from the numerous carnivora which always follow in their train. The migrations of the Lemmings are marked by the devastations they commit along their course, as they come down from the borders of the Frozen Ocean to the valleys of Lapland and Norway; but their migrations are not periodical.

SECTION II.

DISTRIBUTION OF THE FAUNAS.

413. We have stated that all the faunas of the globe may be divided into three groups, corresponding to as many great climatal divisions, namely, the glacial or arctic, the temperate and the tropical faunas. These three divisions appertain to both hemispheres, as we recede from the equator towards the north or south poles. It will hereafter be shown that the tropical and temperate faunas may be again divided into several zoölogical provinces, depending on longitude or on the peculiar configuration of the continents.

414. No continent is better calculated to give a correct idea of distribution into faunas, as determined by climate, than the continent of America; extending as it does across both hemispheres, and embracing all latitudes, so that all climates are represented upon it, as shown by the chart on the following page.

415. Let a traveller embark at Iceland, which is situated on the borders of the polar circle, with a view to observe, in a zoölogical aspect, the principal points along the eastern shore of America. The result of his observation will be very much as follows. Along the coast of Greenland and Iceland, and also along Baffin's Bay, he will meet with an unvaried fauna, composed throughout of the same animals, which are also for the most part identical with those of the arctic shores of Europe. It will be nearly the same along the coast of Labrador.

416. As he approaches Newfoundland, he will see the landscape, and with it the fauna, assuming a somewhat more varied aspect. To the wide and naked or turfy plains of the boreal regions succeed forests, in which he will find

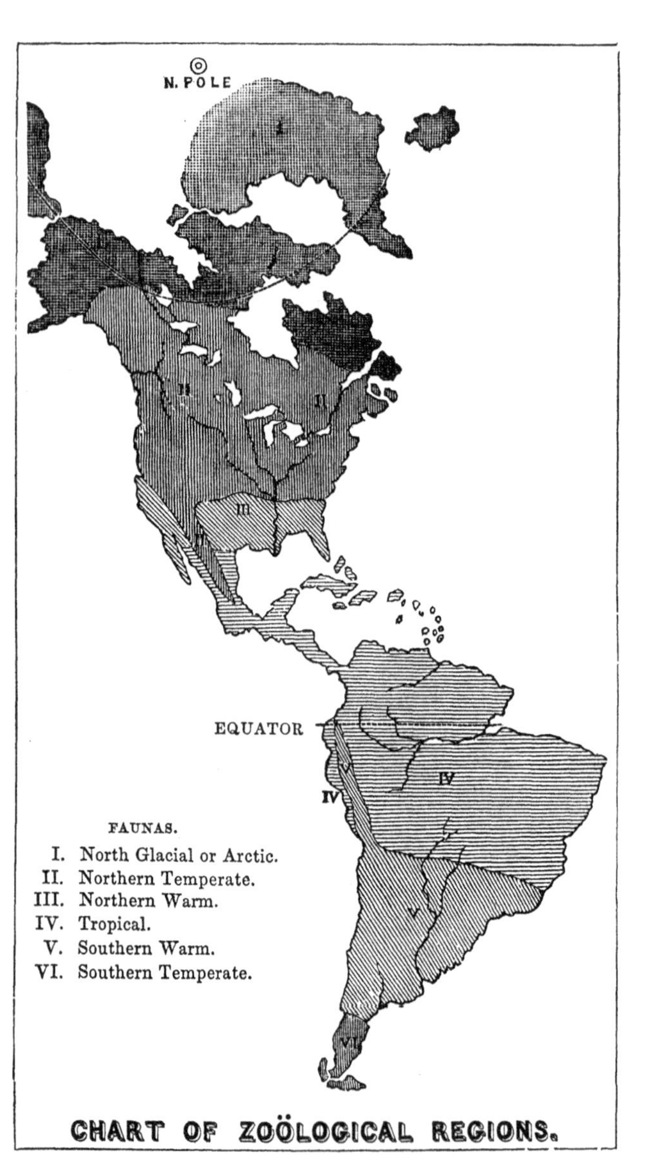

CHART OF ZOÖLOGICAL REGIONS.

various animals which dwell only in forests. Here the temperate fauna commences. Still the number of species is not yet very considerable ; but as he advances southward, along the coasts of Nova Scotia and New England, he finds new species gradually introduced, while those of the colder regions diminish, and at length entirely disappear, some few accidental or periodical visiters excepted, who wander, during winter, as far south as the Carolinas.

417. But it is after having passed the boundaries of the United States, among the Antilles, and more especially on the southern continent, along the shores of the Orinoco and the Amazon, that our traveller will be forcibly struck with the astonishing variety of the animals which people the forests, the prairies, the rivers, and the sea-shores, most of which he will also find to be different from those of the northern continent. By this extraordinary richness of new forms, he will become sensible that he is now in the domain of the tropical fauna.

418. Let him still travel on beyond the equator towards the tropic of Capricorn, and he will again find the scene change as he enters the regions where the sun casts his rays more obliquely, and where the contrast of the seasons is more marked. The vegetation will be less luxuriant; the palms will have disappeared to make place for other trees; the animals will be less varied, and the whole picture will recall to him, in some measure, what he witnessed in the United States. He will again find himself in the temperate region, and this he will trace on, till he arrives at the extremity of the continent, the fauna and the flora becoming more and more impoverished as he approaches Cape Horn.

419. Finally, we know that there is a continent around the South Pole. Although we have as yet but very imperfect notions respecting the animals of this inhospitable clime, still, the few which have already been observed there present

a close analogy to those of the arctic region. It is another glacial fauna, namely, the antarctic. Having thus sketched the general divisions of the faunas, it remains to point out the principal features of each of them.

420. I. Arctic Fauna. — The predominant feature of the Arctic Fauna is its uniformity. The species are few in number; but, on the other hand, the number of individuals is immense. We need only refer to the clouds of birds which hover upon the islands and shores of the North; the shoals of fishes, the salmon among others, which throng the coasts of Greenland, Iceland, and Hudson's Bay. There is great uniformity, also, in the form and color of these animals. Not a single bird of brilliant plumage is found, and few fishes with varied hues. Their forms are regular, and their tints as dusky as the northern heavens. The most conspicuous animals are the white-bear, the moose, the reindeer, the musk-ox, the white-fox, the polar-hare, the lemming, and various Seals; but the most important are the Whales, which, it is to be remarked, rank lowest of all the Mammals. Among the Birds may be enumerated some sea-eagles and a few Waders, while the great majority are aquatic species, such as gulls, cormorants, divers, petrels, ducks, geese, gannets, &c., all belonging to the lowest orders of Birds. Reptiles are altogether wanting. The Articulata are represented by numerous marine worms, and by minute crustaceans of the orders Isopoda and Amphipoda. Insects are rare, and of inferior types. Of the type of Mollusks, there are Acephala, particularly Tunicata, fewer Gasteropods, and very few Cephalopods. Among the Radiata are a great number of jelly-fishes, particularly the Beröe; and to conclude with the Echinoderms, there are several star-fishes and Echini, but few Holothuriæ. The class of Polypi is very scantily represented, and those producing stony corals are entirely wanting.

421. This assemblage of animals is evidently inferior to that of other faunas, especially to those of the tropics. Not that there is a deficiency of animal life; for if the species are less numerous, there is a compensation in the multitude of individuals, and, also, in this other very significant fact, that the largest of all animals, the whales, belong to this fauna.

422. It has already been said, (400,) that the arctic fauna of the three continents is the same; its southern limit, however, is not a regular line. It does not correspond precisely with the polar circle, but rather to the isothermal zero; that is, the line where the average temperature of the year is at 32° of Fahrenheit. The course of this line presents numerous undulations. In general, it may be said to coincide with the northern limit of trees, so that it terminates where forest vegetation succeeds the vast arid plains, the barrens of North America, or the *tundras* of the Samoyedes. The uniformity of these plains involves a corresponding uniformity of plants and animals. On the North American continent it extends much farther southward on the eastern shore than on the western. From the peninsula of Alashka, it bends northwards towards the Mackenzie, then descends again towards the Bear Lake, and comes down nearly to the northern shore of Newfoundland.

423. II. Temperate Faunas. — The faunas of the temperate regions of the northern hemisphere are much more varied than that of the arctic zone. Instead of consisting mainly of aquatic tribes, we have a considerable number of terrestrial animals, of graceful form, animated appearance, and varied colors, though less brilliant than those found in tropical regions. Those parts of the country covered with forests especially swarm with insects, which become the food of other animals; worms and terrestrial and fluviatile mollusks are also abundant.

424. Still, the climate is not sufficiently warm over the whole extent of this zone to allow the trees to retain their foliage throughout the year. At its northern margin, the leaves, excepting those of the pines and spruces, fall, on the approach of the cold season, and vegetation is arrested for a longer or shorter period. Insects retire, and the animals which live upon them no longer find nourishment, and are obliged to migrate to warmer regions, on the borders of the tropics, where, amid the ever-verdant vegetation, they find the means of subsistence.

425. Some of the herbivorous Mammals, the Bats, and the reptiles which feed on insects, pass the winter in a state of torpor, from which they awake in spring. Others retire into dens, and live on the provisions they have stored up during the warm season. The Carnivora, the Ruminants, and the most active portion of the Rodents, are the only animals that do not change either their abode or their habits. The fauna of the temperate zone thus presents an ever-changing picture, which may be considered as one of its most important features, since these changes recur with equal constancy in the Old and the New World.

426. Taking the contrast of the vegetation as a basis, and the consequent changes of habit imposed upon the denizens of the forests, the temperate fauna has been divided into two regions; a northern one, where the trees, except the pines, drop their leaves in winter, and a southern one, where they are evergreen. Now, as the limit of the former, that of the deciduous trees, coincides, in general, with the limit of the pines, it may be said that the cold region of the temperate fauna extends as far as the pines. In the United States this coincidence is not so marked as in other regions, inasmuch as the pines along the Atlantic coast extend into Florida, while they do not prevail in the Western States; but we may consider as belonging to the southern portion

of the temperate region that part of the country south of the latitude where the Palmetto or Cabbage-tree (*Chamærops*) commences, namely, all the States to the south of North Carolina; while the States to the north of this limit belong to the northern portion of the temperate region.

427. This division into two zones is supported by observations made on the maritime faunas of the Atlantic coast The line of separation between them, however, being influenced by the Gulf Stream, is considerably farther to the north, namely, at Cape Cod; although there is also another decided limitation of the marine animals at a point nearly coinciding with the line of demarkation above mentioned, namely, at Cape Hatteras. It has been observed that of one hundred and ninety-seven Mollusks inhabiting the coast of New England, fifty do not pass to the north of Cape Cod, and eighty-three do not pass to the south of it; only sixty-four being common to both sides of the Cape. A similar limitation of the range of Fishes has been noticed by Dr. Storer; and Dr. Holbrook has found the Fishes of South Carolina to be different from those of Florida and the West Indies. In Europe, the northern part of the temperate region extends to the Pyrenees and the Alps; and its southern portion consists of the basin of the Mediterranean, together with the northern part of Africa, as far as the desert of Sahara.

428. A peculiar characteristic of the faunas of the temperate regions in the northern hemisphere, when contrasted with those of the southern, is the great similarity of the prevailing types on both continents. Notwithstanding the immense extent of country embraced, the same stamp is every where exhibited. Generally, the same families, frequently the same genera, represented by different species, are found. There are even a few species of terrestrial animals regarded as identical on the continents of Europe and

America; but their supposed number is constantly diminished, as more accurate observations are made. The predominant types among the mammals are the bison, deer, ox, horse, hog, numerous rodents, especially squirrels and hares, nearly all the insectivora, weasels, martens, wolves, foxes, wildcats, &c. On the other hand, there are no Edentata and no Quadrumana, with the exception of some monkeys, on the two slopes of the Atlas and in Japan. Among Birds, there is a multitude of climbers, passerine, gallinaceous, and many rapacious birds. Of Reptiles, there are lizards and tortoises of small or medium size, serpents, and many batrachians, but no crocodiles. Of fishes, there is the trout family, the cyprinoids, the sturgeons, the pikes, the cod, and especially the great family of Herrings and Scomberoids, to which latter belong the mackerel and the tunny. All classes of the Mollusks are represented; though the cephalopods are less numerous than in the torrid zone. There is an infinite number of Articulata of every type, as well as numerous Polyps, though the corals proper do not yet appear abundantly.

429. On each of the two continents of Europe and America there is a certain number of species, which extend from one extreme of the temperate zone to the other. Such, for example, are the deer, the bison, the cougar, the flying-squirrel, numerous birds of prey, several tortoises, and the rattlesnake, in America. In Europe, the brown bear, wolf, swallow, and many birds of prey. Some species have a still wider range, like the ermine, which is found from Behring's Straits to the Himalaya Mountains, that is to say, from the coldest regions of the arctic zone to the southern confines of the temperate zone. It is the same with the muskrat, which is found from the mouth of Mackenzie's River to Florida. The field-mouse has an equal range in Europe. Other species, on the contrary, are limited to one region.

The Canadian elk is confined to the northern portion of the fauna; while the prairie wolf, the fox-squirrel, the Bassaris, and numerous birds, never leave the southern portion.*

430. In America, as in the Old World, the temperate fauna is further subdivided into several districts, which may be regarded as so many zoölogical provinces, in each of which there is a certain number of animals differing from those in the others, though very closely allied. Temperate America presents us with a striking example in this respect. We have, on the one hand:

1st. The fauna of the United States properly so called, on this side of the Rocky Mountains.

2d. The fauna of Oregon and California, beyond those mountains.

Though there are some animals which traverse the chain of the Rocky Mountains, and are found in the prairies of the Missouri as well as on the banks of the Columbia, as, for example, the Rocky Mountain deer, (*Antilope furcifer*,) yet, if we regard the whole assemblage of animals, they are found to differ entirely. Thus, the rodents, part of the ruminants, the insects, and all the mollusks, belong to distinct species.

431. The faunas or zoölogical provinces of the Old World which correspond to these are:

---

* The types which are peculiar to temperate America, and are not found in Europe, are the Opossum, several genera of Insectivora, among them the shrew-mole (*Scalops aquaticus*) and the star-nose mole, (*Condylura cristata*,) which replaces the Mygale of the Old World; several genera of rodents, especially the muskrat. Among the types characteristic of America must also be reckoned the snapping-turtle among the tortoises; the Menobranchus and Menopoma, among the Salamanders; the Gar-pike and Amia among the fishes; and finally, among the Crustacea, the Limulus. Among the types which are wanting in temperate America, and which are found in Europe, may be cited the horse, the wild boar, and the true mouse. All the species of domestic mice which live in America have been brought from the Old World.

1st. The fauna of Europe, which is very closely related to that of the United States proper.

2d. The fauna of Siberia, separated from the fauna of Europe by the Ural Mountains.

3d. The fauna of the Asiatic table-land, which, from what is as yet known of it, appears to be quite distinct.

4th. The fauna of China and Japan, which is analogous to that of Europe in the Birds, and to that of the United States in the Reptiles — as it it also in the flora.

Lastly, it is in the temperate zone of the northern hemisphere that we meet with the most striking example of those local faunas which have been mentioned above. Such, for example, is the fauna of the Caspian Sea, of the steppes of Tartary, and of the Western prairies.

432. The faunas of the southern temperate regions differ from those of the tropics as much as the northern temperate faunas do; and, like them also, may be distinguished into two provinces, the colder of which embraces Patagonia. But besides differing from the tropical faunas, they are also quite unlike each other on the different continents. Instead of that general resemblance, that family likeness which we have noticed between all the faunas of the temperate zone of the northern hemisphere, we find here the most complete contrasts. Each of the three continental peninsulas which jut out southerly into the ocean represents, in some sense, a separate world. The animals of South America, beyond the tropic of Capricorn, are in all respects different from those at the southern extremity of Africa. The hyenas, wild-boars, and rhinoceroses of the Cape of Good Hope have no analogues on the American continent; and the difference is equally great between the birds, reptiles and fishes, insects and mollusks. Among the most characteristic animals of the southern extremity of America are peculiar species of seals, and especially, among aquatic birds, the penguins.

433. New Holland, with its marsupial mammals, with which are associated insects and mollusks no less singular furnishes a fauna still more peculiar, and which has no similarity to those of any of the adjacent countries. In the seas of that continent, where every thing is so strange, we find the curious shark, with paved teeth and spines on the back, (*Cestracion Philippii*,) the only living representative of a family so numerous in former zoölogical ages. But a most remarkable feature of this fauna is, that the same types prevail over the whole continent, in its temperate as well as its tropical portions, the species only being different at different localities.

434. Tropical Faunas. — The tropical faunas are distinguished, on all the continents, by the immense variety of animals which they comprise, not less than by the brilliancy of their dress. All the principal types of animals are represented, and all contain numerous genera and species. We need only refer to the tribe of humming-birds, which numbers not less than **300** species. It is very important to notice, that here are concentrated the most perfect, as well as the oddest, types of all the classes of the Animal Kingdom. The tropical region is the only one occupied by the Quadrumana, the herbivorous bats, the great pachydermata, such as the elephant, the hippopotamus, and the tapir, and the whole family of Edentata. Here also are found the largest of the cat tribe, the lion and tiger. Among the Birds we may mention the parrots and toucans, as essentially tropical; among the Reptiles, the largest crocodiles, and gigantic tortoises; and finally, among the articulated animals, an immense variety of the most beautiful insects. The marine animals, as a whole, are equally superior to those of other regions; the seas teem with crustaceans and numerous cephalopods, together with an infinite variety of gasteropods and acephala. The Echinoderms there attain a magnitude

and variety elsewhere unknown; and lastly, the Polyps there display an activity of which the other zones present no example. Whole groups of islands are surrounded with coral reefs formed by those little animals.

435. The variety of the tropical fauna is further enriched by the circumstance that each continent furnishes new and peculiar forms. Sometimes whole types are limited to one continent, as the sloth, the toucans, and the humming-birds to America, the giraffe and hippopotamus to Africa; and again animals of the same group have different characteristics, according as they are found on different continents. Thus, the monkeys of America have flat and widely separated nostrils, thirty-six teeth, and generally a long, prehensile tail. The monkeys of the Old World, on the contrary, have nostrils close together, only thirty-two teeth, and not one of them has a prehensile tail.

436. But these differences, however important they may appear at first glance, are subordinate to more important characters, which establish a certain general affinity between all the faunas of the tropics. Such, for example, is the fact that the quadrumana are limited, on all the continents, to the warmest regions; and never, or but rarely, penetrate into the temperate zone. This limitation is a natural consequence of the distribution of the palms; for as these trees, which constitute the ruling feature of the flora of the tropics, furnish, to a great extent, the food of the monkeys on both continents, we have only to trace the limits of the palms, to have a pretty accurate indication of the extent of the tropical faunas on all three continents.

437. Several well-marked faunas may be distinguished in the tropical part of the American continent, namely:

1. The fauna of Brazil, characterized by its gigantic reptiles, its monkeys, its Edentata, its tapir, its humming-birds, and its astonishing variety of insects.

2. The fauna of the western slope of the Andes, comprising Chili and Peru; and distinguished by its Llamas, vicuñas, and birds, which differ from those of the basin of the Amazon, as also do the insects and mollusks.

3. The fauna of the Antilles and the Gulf of Mexico. This is especially characterized by its marine animals, among which the Manatée is particularly remarkable; an infinite variety of singular fishes, embracing a large number of Plectognaths; also Mollusks, and Radiata of peculiar species. It is in this zone that the *Pentacrinus caput-medusæ* is found, the only representative, in the existing creation, of a family so numerous in ancient epochs, the Crinoidea with a jointed stem.

The limits of the fauna of Central America cannot yet be well defined, from want of sufficient knowledge of the animals which inhabit those regions.

438. The tropical zone of Africa is distinguished by a striking uniformity in the distribution of the animals, which corresponds to the uniformity of the structure and contour of that continent. Its most characteristic species are spread over the whole extent of the tropics: thus, the giraffe is met with from Upper Egypt to the Cape of Good Hope. The hippopotamus is found at the same time in the Nile, the Niger, and Orange River. This wide range is the more significant as it also relates to herbivorous animals, and thus supposes conditions of vegetation very similar, over wide countries. Some forms are, nevertheless, circumscribed within narrow districts; and there are marked differences between the animals of the eastern and western shores. Among the remarkable species of the African torrid region are the baboons, the African elephant, the crocodile of the Nile, a vast number of Antelopes, and especially two species of Orang-outang, the Chimpanzée and the Engeena, a large and remarkable animal, only recently described. The fishes of the Nile have a tropical character, as well as the animals

of Arabia, which are more allied to those of Africa than to those of Asia.

439. The tropical fauna of Asia, comprising the two peninsulas of India and the Isles of Sunda, is not less marked. It is the country of the gibbons, the red orang, the royal tiger, the gavial, and a multitude of peculiar birds. Among the fishes, the family of Chetodons is most numerously represented. Here also are found those curious spiny fishes, whose intricate gills suggested the name Labyrinthici, by which they are known. Fishes with tufted gills are more numerous here than in other seas. The insects and mollusks are no less strongly characterized. Among others is the nautilus, the only living representative of the great family of large, chambered-shells which prevailed so extensively over other types, in former geological ages.

440. The large Island of Madagascar has its peculiar fauna, characterized by its makis and its curious rodents. It is also the habitat of the Aya-aya. Polynesia, exclusive of New Holland, furnishes a number of very curious animals, which are not found on the Asiatic continent. Such are the herbivorous bats, and the Galeopithecus or flying Maki. The Galapago islands, only a few hundred miles from the coast of Peru, have a fauna exclusively their own, among which gigantic land-tortoises are particularly characteristic.

SECTION III.

CONCLUSIONS.

441. From the survey we have thus made of the distribution of the Animal Kingdom, it follows:

1st. Each grand division of the globe has animals which are either wholly or for the most part peculiar to it. These groups of animals constitute the faunas of different regions.

2d. The diversity of faunas is not in proportion to the distance which separates them. Very similar faunas are found at great distances apart; as, for example, the fauna of Europe and that of the United States, which yet are separated by a wide ocean. Others, on the contrary, differ considerably, though at comparatively short distances; as the fauna of the East Indies and the Sunda Islands, and that of New Holland; or the fauna of Labrador and that of New England.

3d. There is a direct relation between the richness of a fauna and the climate. The tropical faunas contain a much larger number of more perfect animals than those of the temperate and polar regions.

4th. There is a no less striking relation between the fauna and flora, the limit of the former being oftentimes determined, so far as terrestrial animals are concerned, by the extent of the latter.

442. Animals are endowed with instincts and faculties corresponding to the physical character of the countries they inhabit, and which would be of no service to them under other circumstances. The monkey, which is a frugivorous animal, is organized for living on the trees from which he obtains his food. The reindeer, on the contrary, whose food consists of lichens, lives in cold regions. The latter would be quite out of place in the torrid zone, and the monkey would perish with hunger in the polar regions. Animals which store up provisions are all peculiar to temperate or cold climates. Their instincts would be uncalled for in tropical regions, where the vegetation presents the herbivora with an abundant supply of food at all times.

443. However intimately the climate of a country seems to be allied with the peculiar character of its fauna, we are not to conclude that the one is the consequence of the other. The differences which are observed between the animals of

different faunas are no more to be ascribed to the influences of climate, than their organization is to the influence of the physical forces of nature. If it were so, we should necessarily find all animals precisely similar, when placed under the same circumstances. We shall find, by the study of the different groups in detail, that certain species, though very nearly alike, are nevertheless distinct in two different faunas. Between the animals of the temperate zone of Europe, and those of the United States, there is similarity but not identity; and the particulars in which they differ, though apparently trifling, are yet constant.

444. Fully to appreciate the value of these differences, it is often requisite to know all the species of a genus or of a family. It is not uncommon to find, upon such an examination, that there is the closest resemblance between species that dwell far apart from each other, while species of the same genus, that live side by side, are widely different. This may be illustrated by a single example. The Menopoma, Siren, Amphiuma, Axolotl, and the Menobranchus, are Batrachians which inhabit the rivers and lakes of the United States and Mexico. They are very similar in external form, yet differ in the fact that some of them have external gills at the sides of the head, in which others are deficient; that some have five legs, while others are only provided with two; and also in having either two or four legs. Hence we might be tempted to refer them to different types, did we not know intermediate animals, completing the series, namely, the Proteus and Megalobatrachus. Now, the former exists only in the subterranean lakes of Austria, and the latter in Japan. The connection in this case is consequently established by means of species which inhabit continents widely distant from each other.

445. Neither the distribution of animals, therefore, any more than their organization, can be the effect of external

influences. We must, on the contrary, see in it the realization of a plan wisely designed, the work of a Supreme Intelligence who created, at the beginning, each species of animal at the place, and for the place, which it inhabits. To each species has been assigned a limit which it has no disposition to overstep, so long as it remains in a wild state. Only those animals which have been subjected to the yoke of man, or whose subsistence is dependent on man's social habits, are exceptions to this rule.

446. As the human race has extended over the surface of the earth, man has more or less modified the animal population of different regions, either by exterminating certain species, or by introducing others with which he desires to be more intimately associated — the domestic animals. Thus, the dog is found wherever we know of the presence of man. The horse, originally from Asia, was introduced into America by the Spaniards; where it has thriven so well, that it is found wild, in innumerable herds, over the Pampas of South America, and the prairies of the West. In like manner, the domestic ox became wild in South America. Many less welcome animals have followed man in his peregrinations; as, for example, the rat and the mouse, as well as a multitude of insects, such as the house-fly, the cockroach, and others which are attached to certain species of plants, as the white butterfly, the Hessian fly, &c. The honey-bee, also, has been imported from Europe.

447. Among the species which have disappeared, under the influence of man, we may mention the Dodo, a peculiar species of bird which once inhabited the Mauritius, some remains of which are preserved in the British and Ashmolean Museums; also a large cetacean of the north, (*Rytina Stelleri*,) formerly inhabiting the coasts of Behring's Straits, and which has not been seen since 1768. According to all appearances, we must also count among these the great stag, the skeleton and horns of which have been found buried in the peat-bogs of Ireland. There are also many species of animals whose numbers are daily diminishing, and whose extinction may be foreseen ; as the Canada deer, (*Wapiti*,) the Ibex of the Alps, the Lämmergeyer, the bison, the beaver, the wild turkey, &c.

448. Other causes may also contribute towards dispersing animals beyond their natural limits. Thus, the sea-weeds are carried about by marine currents, and are frequently met with far from shore, thronged with little crustaceans, which are in this manner transported to great distances from the place of their birth. The drift wood which the Gulf Stream floats from the Gulf of Mexico even to the western shores of Europe, is frequently perforated by the larvæ of insects, and may, probably, serve as depositories for the eggs of fishes, crustacea, and mollusks. It is possible, also, that aquatic birds may contribute in some measure to the diffusion of some species of fishes and mollusks, either by the eggs becoming attached to their feet, or by means of those which they evacuate undigested, after having transported them to considerable distances. Still, all these circumstances exercise but a very feeble influence upon the distribution of species in general ; and each country, none the less, preserves its peculiar physiognomy, so far as its animals are concerned.

449. There is only one way to account for the distribution of animals as we find them, namely, to suppose that they are *autochthonoi*, that is to say, that they originated like plants, on the soil where they are found. In order to explain the particular distribution of many animals, we are even led to admit that they must have been created at several points of the same zone ; an inference which we must make from the distribution of aquatic animals, especially that of Fishes. If we examine the fishes of the different

rivers of the United States, peculiar species will be found in each basin, associated with others which are common to several basins. Thus, the Delaware River contains species not found in the Hudson. But, on the other hand, the pickerel is found in both. Now, if all animals originated at one point, and from a single stock, the pickerel must have passed from the Delaware to the Hudson, or *vice versa*, which it could only have done by passing along the sea-shore, or by leaping over large spaces of *terra firma;* that is to say, in both cases it would be necessary to do violence to its organization. Now, such a supposition is in direct opposition to the immutability of the laws of Nature.

450. We shall hereafter see that the same laws of distribution are not limited to the actual creation only, but that they have also ruled the creations of former geological epochs, and that the fossil species have lived and died, most of them, at the place where their remains are found.

451. Even Man, although a cosmopolite, is subject, in a certain sense, to this law of limitation. While he is every where the one identical species, yet several races, marked by certain peculiarities of features, are recognized; such as the Caucasian, Mongolian, and African races, of which we are hereafter to speak. And it is not a little remarkable, that the abiding places of these several races correspond very nearly with some of the great zoölogical regions. Thus we have a northern race, comprising the Samoyedes in Asia, the Laplanders in Europe, and the Esquimaux in America, corresponding to the arctic fauna, (**400**,) and, like it, identical on the three continents, having for its southern limit the region of trees, (422.) In Africa, we have the Hottentot and Negro races, in the south and central portions respectively, while the people of northern Africa are allied to their neighbors in Europe; just as we have seen to be the case with the zoölogical fauna in general,

(403.) The inhabitants of New Holland, like its animals, are the most grotesque and uncouth of all races, (433.)

452. The same parallelism holds good elsewhere, though not always in so remarkable a degree. In America, especially, while the aboriginal race is as well distinguished from other races as is its flora, the minor divisions are not so decided. Indeed, the facilities, or we might sometimes rather say necessities, arising from the varied supplies of animal and vegetable food in the several regions, might be expected to involve, with his corresponding customs and modes of life, a difference in the physical constitution of man, which would contribute to augment any primeval differences. It could not indeed be expected, that a people constantly subjected to cold, like the people of the North, and living almost exclusively on fish, which is not to be obtained without great toil and peril, should present the same characteristics, either bodily or mental, as those who idly regale on the spontaneous bounties of tropical vegetation.

CHAPTER FOURTEENTH.

GEOLOGICAL SUCCESSION OF ANIMALS; OR, THEIR DISTRIBUTION IN TIME.

SECTION I.

STRUCTURE OF THE EARTH'S CRUST.

453. THE records of the Bible, as well as human tradition, teach us that man and the animals associated with him were created by the word of God; "the Lord made heaven and earth, the sea, and all that in them is;" and this truth is confirmed by the revelations of science, which unequivocally indicate the direct interventions of creative power.

454. But man and the animals which now surround him are not the only kinds which have had a being. The surface of our planet, anterior to their appearance, was not a desert. There are, scattered through the crust of the earth, numerous animal and vegetable remains, which show that the earth had been repeatedly supplied with, and long inhabited by, animals and plants altogether different from those now living.

455. In general, their hard parts are the only relics of them which have been preserved, such as the skeleton and teeth of Vertebrates; the shells of the Mollusks and Radiata; the shields of the Crustaceans, and sometimes the wing-cases of Insects. Most frequently they have lost their original

chemical composition, and are changed into stone; and hence the name of *petrifactions* or *fossils*, under which latter term are comprehended all the organized bodies of former epochs, obtained from the earth's crust. Others have entirely disappeared, leaving only their forms and sculpture impressed upon the rocks.

456. The study of these remains and of their position in the rocks constitutes PALEONTOLOGY; one of the most essential branches of Zoölogy. Their geological distribution, or the order of their successive appearance, namely, *the distribution of animals in time,* is of no less importance than the geographical distribution of living animals, their distribution *in space,* of which we have treated in the preceding chapter. To obtain an idea of the successive creations, and of the stupendous length of time they have required, it is necessary to sketch the principal outlines of Geology.

457. The rocks* which compose the crust of our globe are of two kinds:

1. The *Massive Rocks*, called also *Plutonic* or *Igneous Rocks*, which lie beneath all the others, or have sometimes been forced up through them, from beneath. They were once in a melted state, like the lava of the present epoch, and on cooling at the surface formed the original crust of the globe, the granite, and later porphyry, basalt, &c.

2. The *Sedimentary* or *Stratified Rocks*, called also *Neptunic Rocks,* which have been deposited in water, in the same manner as modern seas and lakes deposit sand and mud on their shores, or at the bottom.

458. These sediments have been derived partly from the disintegration of the older rocks, and partly from the decay of plants and animals. The materials being disposed in

* Rocks, in a geological sense, include all the materials of the earth, the loose soil and gravel, as well as the firm rock.

layers or strata, have become, as they hardened, limestones, slates, marls, or grits, according to their chemical and mechanical composition, and contain the remains of the animals and plants which were scattered through the waters.*

459. The different strata, when undisturbed, are arranged one above the other in a horizontal manner, like the leaves of a book, the lowest being the oldest. In consequence of the commotions which the crust of the globe has undergone, the strata have been ruptured, and many points of the surface have been elevated to great heights, in the form of mountains; and hence it is that fossils are sometimes found at the summit of the highest mountains, though the rocks containing them were originally formed at the bottom of the sea. But even when folded, or partly broken, their relative age may still be determined by an examination of the ends of the upturned strata, where they appear or crop out in succession, at the surface, or on the slopes of mountains, as seen in the diagram, (Fig. 154.)

460. The sedimentary rocks are the only ones which have been found to contain animal and vegetable remains. These are found imbedded in the rock, just as we should find them in the mud now deposited at the bottom of the sea, if laid dry. The strata containing fossils are numerous. The comparison and detailed study of them belongs to Geology, of

---

* Underneath the deepest strata containing fossils, between these and the Plutonic rocks, are generally found very extensive layers of slates without fossils, (gneiss, mica-slate, talcose-slate,) though stratified, and known to the geologist under the name of *Metamorphic Rocks*, (Fig. 154, *M*,) being probably sedimentary rocks, which have undergone considerable changes. The Plutonic rocks, as well as the metamorphic rocks, are not always confined to the lower levels, but they are often seen rising to considerable heights, and forming many of the loftiest peaks of the globe. The former also penetrate, in many cases, like veins, through the whole mass of the stratified and metamorphic layers, and expand at the surface; as is the case with the trap dykes, and as lava streams actually do at the present era, (Fig. 154, *T. L.*)

which Paleontology forms an essential part. A group of strata extending over a certain geographical extent, all of which contain some fossils in common, no matter what may be the chemical character of the rock, whether it be limestone, sand, or clay, is termed a geological *Formation.* Thus, the coal beds, with the intervening slates and grits, and the masses of limestone, between which they often lie, constitute but one formation — the carboniferous formation.

461. Among the stratified rocks we distinguish ten principal Formations, each of which indicates an entirely new era in the earth's history; while each of the layers which compose a formation indicates but some partial revolution. Proceeding from below upwards, they are as follows, as indicated in the cut, and also in the lower diagram on the Frontispiece.

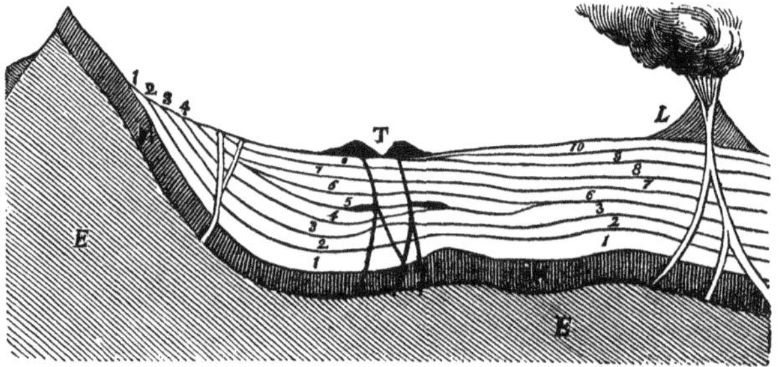

Fig. 154.

1st. The *Lower Silurian.* This is a most extensive formation, no less than eight stages of which have been made out by Geologists in North America, composed of various limestones and sandstones.*

* 1. Potsdam Sandstone; 2. Calciferous Sandstone; 3. Chazy Lime stone; 4. Bird's-eye Limestone; 5. Black River Limestone; 6. Trenton Limestone; 7. Utica Slate; 8. Hudson River Group; being all found in the western parts of the United States.

2d. The *Upper Silurian.* It is also a very extensive formation, since about ten stages of it are found in the State of New York.*

3d. The *Devonian*, including in North America no less than eleven stages.† It occurs also in Russia and Scotland, where it was first made out as a peculiar formation.

4th. The *Carboniferous Formation*, consisting of three grand divisions.‡

5th. The *Trias*, or *Saliferous Formation*, which, containing the richest deposits of Salt on the continent of Europe, comprises three stages,§ to one of which the Sandstone of the Connecticut valley belongs.

6th. The *Oölitic Formation*, only faint traces of which exist on the continent of America. It comprises at least four distinct stages.||

7th. The *Cretaceous*, or *Chalk Formation*, of which three principal stages have been recognized, two of which are feebly represented in this country, in the Southern and Middle States.

8th. The *Lower Tertiary*, or *Eocene*, very abundant in the Southern States of the Union, and to which belong the coarse limestone of Paris, and the London clay in England.

---

* 1. Oneida Conglomerate; 2. Medina Sandstone; 3. Clinton Group; 4. Niagara Group; 5. Onondaga Salt Group; 6. Water Limestone; 7. Pentamerus Limestone; 8. Delthyris Shaly Limestone; 9. Encrinal Limestone; 10. Upper Pentamerus Limestone.

† 1. Oriskany Sandstone; 2. Cauda-Galli Grit; 3. Onondaga Limestone; 4. Corniferous Limestone; 5. Marcellus Shale; 6. Hamilton Group; 7. Tully Limestone; 8. Genesee Slate; 9. Portage Group; 10. Chemung Group; 11. Old Red Sandstone.

‡ 1. The Permian, extensively developed in Russia, especially in the government of Perm; 2. The coal measures, containing the rich deposits of coal in the Old and New World; 3. The Magnesian Limestone of England.

§ 1. New Red Sandstone; 2. Muschelkalk; 3. Keuper.

|| 1. The Lias; 2. The Lower Oölite; 3. The Middle Oölite; 4 The Upper Oölite.

9th. The *Upper Tertiary*, or *Miocene* and *Pleiocene*, found also in the United States, as far north as Martha s Vineyard and Nantucket, and very extensive in Southern Europe, as well as in South America.

10th. The *Drift*, forming the most superficial deposits, and extending over a large portion of the northern countries in both hemispheres.

We have thus more than forty distinct layers already made out, each of which marks a distinct epoch in the earth's history, indicating a more or less extensive and important change in the condition of its surface.

462. All the formations are not every where found, or are not developed to the same extent, in all places. So it is with the several strata of which they are composed. In other words, the layers of the earth's crust are not continuous throughout, like the coats of an onion. There is no place on the globe where, if it were possible to bore down to its centre, all the strata would be found. It is easy to understand how this must be so. Since irregularities in the distribution of water upon the solid crust have, necessarily, always existed to a certain extent, portions of the earth's surface must have been left dry at every epoch of its history, gradually forming large islands and continents, as the changes were multiplied. And since the rocks were formed by the subsidence of sediment in water, no rocks would be formed except in regions covered by water; they would be thickest at the parts where most sediment was deposited, and gradually thin out towards their circumference. We may therefore infer, that all those portions of the earth's surface which are destitute of a certain formation were dry land, during that epoch of the earth's history to which such formation relates, excepting, indeed, where the rocks have been subsequently removed by the denuding action of water or other causes.

463. Each formation represents an immense period of time, during which the earth was inhabited by successive races of animals and plants, whose remains are often found, in their natural position, in the places where they lived and died, not scattered at random, though sometimes mingled together by currents of water, or other influences, subsequent to the time of their interment. From the manner in which the remains of various species are found associated in the rock, it is easy to determine whether the animals to which these remains belonged lived in the water, or on land, on the beach or in the depths of the ocean, in a warm or in a cold climate. They will be found associated in just the same way as animals are that live under similar influences at the present day.

464. In most geological formations, the number of species of animals and plants found in any locality of given extent, is not below that of the species now living in an area of equal extent and of a similar character; for though, in some deposits, the variety of the animals contained may be less, in others it is greater than that on the present surface. Thus, the coarse limestone in the neighborhood of Paris, which is only one stage of the lower tertiary, contains not less than 1200 species of shells; whereas the species now living in the Mediterranean do not amount to half that number. Similar relations may be pointed out in America. Mr. Hall, one of the geologists of the New York Survey, has described, from the Trenton limestone, (one of the ten stages of the lower Silurian,) 170 species of shells, a number almost equal to that of all the species found now living on the coast of Massachusetts.

465. Nor was the number of individuals less than at present. Whole rocks are entirely formed of animal remains, particularly of corals and shells. So, also, coal is composed of the remains of plants. If we consider the slow-

ness with which corals and shells are formed, it will give us some faint notion of the vast series of ages that must have elapsed in order to allow the formation of those rocks, and their regular deposition, under the water, to so great a thickness. If, as all things combine to prove, this deposition took place in a slow and gradual manner in each formation, we must conclude, that the successive species of animals found in them followed each other at long intervals, and are not the work of a single epoch.

466. It was once believed that animals were successively created in the order of their relative perfection; so that the most ancient formations contained only animals of the lowest grade, such as the Polyps, the Echinoderms, to which succeeded the Mollusks, then the Articulated Animals, and, last of all, the Vertebrates. This theory, however, is now untenable; since fossils belonging to each of the four departments have been found in the fossiliferous deposits of every age. Indeed, we shall see that even in the lower Silurian formation there exist not only Polyps and other Radiata, but also numerous Mollusks, Trilobites, (belonging to the Articulata,) and even Fishes.

SECTION II.

AGES OF NATURE.

467. Each formation, as has been before stated, (460,) contains remains peculiar to itself, which do not extend into the neighboring deposits above or below it. Still there is a connection between the different formations, more strong in proportion to their proximity to each other. Thus, the animal remains of the Chalk, while they differ from those of all other formations, are, nevertheless, much more nearly related

to those of the Oölitic formation, which immediately precedes, than to those of the carboniferous formation, which is much more ancient; and, in the same manner, the fossils of the carboniferous group approach more nearly to those of the Silurian formation than to those of the Tertiary.

468. These relations could not escape the observation of naturalists, and indeed they are of great importance for the true understanding of the development of life at the surface of our earth. And, as in the history of man, several grand periods have been established, under the name of *Ages*, marked by peculiarities in his social and intellectual condition, and illustrated by contemporaneous monuments, so, in the history of the earth, also, are distinguished several great periods, which may be designated as the various *Ages of Nature*, illustrated, in like manner, by their monuments, the fossil remains, which, by certain general traits stamped upon them, clearly indicate the eras to which they belong.

469. We distinguish four Ages of Nature, corresponding to the great geological divisions, namely:

1st. *The Primary* or *Palæozoic Age*, comprising the lower Silurian, the upper Silurian, and the Devonian. During this age there were no air-breathing animals. The fishes were the masters of creation. We may therefore call it the *Reign of Fishes.*

2d. *The Secondary Age*, comprising the carboniferous formation, the Trias, the Oölitic, and the Cretaceous formations. This is the epoch in which air-breathing animals first appear. Reptiles predominate over the other classes, and we may therefore call it the *Reign of Reptiles.*

3d. *The Tertiary Age*, comprising the tertiary formations. During this age, terrestrial mammals, of great size, abound. This is the *Reign of Mammals.*

4th. *The Modern Age*, characterized by the appearance of the most perfect of all created beings. This is the *Reign of Man.*

Let us review each of these four Ages of Nature, with reference to the diagram at the beginning of the volume.

470. THE PALÆOZOIC AGE. *Reign of Fishes.* — The palæozoic fauna, being the most remote from the present epoch, presents the least resemblance to the animals now existing, as will easily be perceived by a glance at the fol-

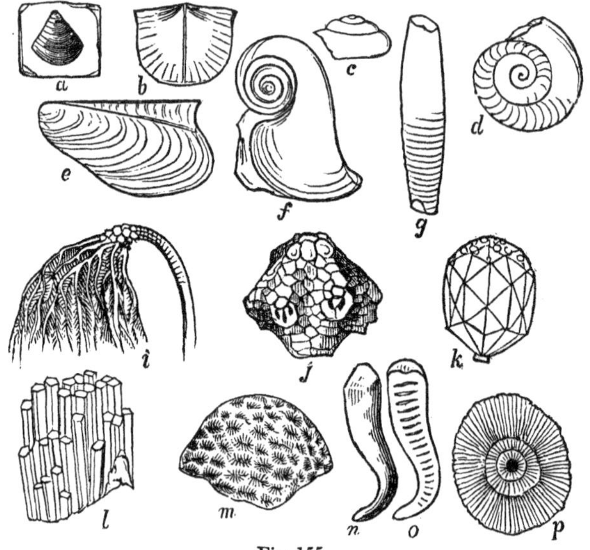

Fig. 155.

lowing sketches, (Fig. 155.) In no other case do we meet with animals of such extraordinary shapes, as in the strata of the Palæozoic age.

471. We have already stated (466) that there are found, in each formation of the primary age, animal remains of all the four great departments, namely, vertebrates, articulata, mollusks, and radiata. We have now to examine to what peculiar classes and families of each department these remains belong, with a view o ascertain if any relation between

the structure of an animal, and the epoch of its first appearance on the earth's surface, may be traced.

472. As a general result of the inquiries hitherto made, it may be stated that the palæozoic animals belong, for the most part, to the lower divisions of the different classes. Thus, of the class of Echinoderms, we find scarcely any but Crinoids, which are the least perfect of the class. We have represented, in the above sketches, several of the most curious forms,* as well as of the Polyps, of which there are some quite peculiar types from the Trenton limestone, and from the Black River limestone.

473. Of the Mollusks, the bivalves or Acephala are numerous, but, for the most part, they belong to the Brachiopoda, that is to say, to the lowest division of the class, including mollusks with unequal valves, having peculiar appendages in the interior. The *Leptæna alternata*, (*b*,) which is found very abundantly in the Trenton limestone, is one of these shells. The only fossils yet found in the Potsdam sandstone, the oldest of all fossiliferous deposits, belong, also, to this family, (*Lingula prima*, *a*.) Besides this, there are also found some bivalves of a less uncommon shape, (*Avicula decussata*, *e*.)

474. The Gasteropods are less abundant; some of them are of a peculiar shape and structure, (*Bucania expansa*, *f*; *Euomphalus hemisphericus*, *c*.) Those more similar to our common marine snails have all an entire aperture; those with a canal being of a more recent epoch.

475. Of the Cephalopods we find some genera not less curious, part of which disappear in the succeeding epochs;

---

* (*i*) *Cyathocrinus ornatissimus*, Hall; (*j*) *Melocrinus Amphora*, Goldf.; (*k*) *Cariocrinus ornatus*, Say; (*l*) *Columnaria alveolata*; (*m*) *Cyathophyllum quadrigeminum*, Goldf.; (*n*, *o*) *Caninia flexuosa*; (*p*) *Chætetes lycoperdon*.

such, in particular, as those of the straight, chambered shells called Orthoceratites, some of which are twelve feet in length, (*Orthoceras fusiforme*, *g*.) There are also found some of a coiled shape, like the Ammonites of the secondary age, but having less complicated partitions, (*Trocholites ammonius*, *d*.) The true cuttle-fishes, which are the highest of the class, are not yet found. On the contrary, the Bryozoa, which have long been considered as polyps, but which, according to all appearances, are mollusks of a very low order, are very numerous in this epoch.

476. The Articulata of the Palæozoic age are mostly Trilobites, animals which evidently belong to the lower order of the Crustaceans, (Fig. 156.) There is an incompleteness and want of development, in the form of their body, that strongly reminds us of the embryo among the crabs. A great many genera have already been discovered.

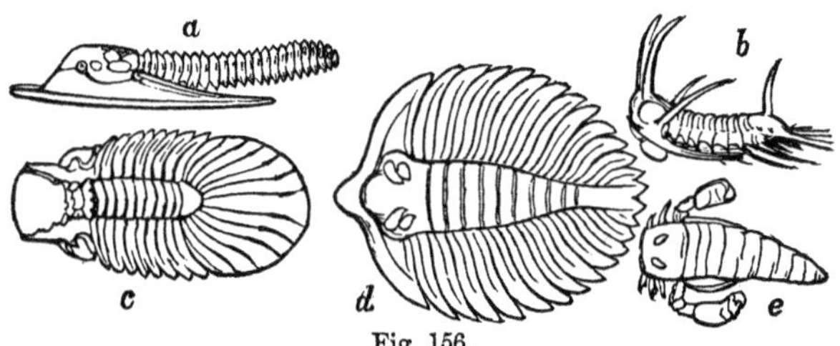

Fig. 156.

We may consider as belonging to the more extraordinary the forms here represented, (*Harpes*, *a*; *Arges*, *b*; *Brontes*, *c*; and *Platynotus*, *d*;) the latter, as well as the *Isotelus*, the largest of all, being peculiar to the Palæozoic deposit of this country. Some others seem more allied to the crustaceans of the following ages, but are nevertheless of a very extraordinary form, as *Eurypterus remipes*, (*e*.) There are a so found, in the Devonian, some very large Entomostraca. The class of Worms is represented only by a few Serpulæ, which are marine worms, surrounded by a solid sheath. The class of Insects is entirely wanting.

477. The inferiority of the earliest inhabitants of our earth appears most striking among the Vertebrates. There are as yet neither reptiles, birds, nor mammals. The fishes, as we have said, are the sole representatives of this division of animals.

478. But the fishes of that early period were not like ours. Some of them had the most extraordinary forms, so that they have been often mistaken for quite different animals; for example, the *Pterichthys*, (*a*,) with its two wing-

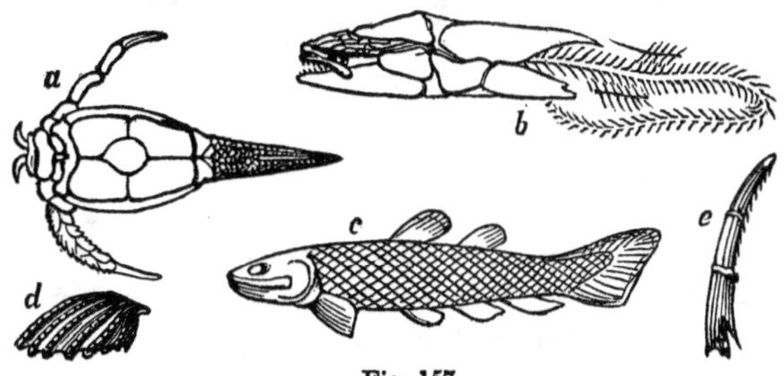

Fig. 157.

like appendages, and also the *Coccosteus* (*b*) of the same deposit, with its large plates covering the head and the anterior part of the body. There are also found remains of shark's spines, (*e*,) as well as palatal bones, (*d*,) the latter of a very peculiar kind. Even those fishes which have a more regular shape, as the *Dipterus*, (*c*,) have not horny scales like our common fishes, but are protected by a coat of bony plates, covered with enamel, like the gar-pikes of the American rivers. Moreover, they all exhibit certain characteristic features, which are very interesting in a physiological point of view. They all have a broad head, and a tail terminating in two unequal lobes. What is still more curious, the best preserved specimens show no indications of the bodies of vertebræ, but merely of their spinous processes; from which it must be inferred that the body of the vertebra was cartilaginous, as it is in our Sturgeons.

479. Recurring to what has been stated on that point, in Chapter Twelfth, we thence conclude, that these ancient fishes were not so fully developed as most of our fishes, being, like the Sturgeon, arrested, as it were, in their development; since we have shown that the Sturgeon, in its organization, agrees, in many respects, with the Cod or Salmon at an early age.

480. Finally, there was, during the Palæozoic age, but little variety among the animals of the different regions of the globe; and this may be readily explained by the peculiar configuration of the earth at that epoch. Great mountains did not then exist; there were neither lofty elevations nor deep depressions. The sea covered the greater part, if not the whole, of the surface of the globe; and the animals which then existed, and whose remains have been preserved, were all, without exception, aquatic animals, breathing by gills. This wide distribution of the waters impressed a very uniform character upon the whole Animal Kingdom. Between the different zones and continents, no such strange contrasts of the different types existed as at the present epoch. The same genera, and often the same species, were found in the seas of America, Europe, Asia, Africa, and New Holland; from which we must conclude that the climate was much more uniform than at the present day. Among the aquatic population, no sound was heard. All creation was then silent.

481. THE SECONDARY AGE. *Reign of Reptiles.* — The Secondary age displays a greater variety of animals as well as plants. The fantastic forms of the Palæozoic age disappear, and in their place we see a greater symmetry of shape. The advance is particularly marked in the series of verte-

brates. Fishes are no longer the sole representatives of that department. Reptiles, Birds, and Mammals successively make their appearance, but Reptiles are preponderant, particularly in the oölitic formation; on which account we have called this the *Reign of Reptiles.*

482. The carboniferous formation is the most ancient of the Secondary age. Its fauna bears, in various respects, a close analogy to that of the Palæozoic epoch, especially in its Trilobites and Mollusks.* Besides these, we meet here with the first air-breathing animals, which are Insects and Scorpions. At the same time, land-plants first make their appearance, namely, ferns of great size, club-mosses, and other fossil plants. This corroborates what has been already said concerning the intimate connection that exists, and from all times has existed, between animals and the land-plants, (399.) The class of Crustaceans has also improved during the epoch of the coal. It is no longer composed exclusively of Trilobites, but the type of horse-shoe crabs also appears, with other gigantic forms. Some of the Mollusks seem also to approach those of the Oölitic period, particularly the Bivalves.

483. In the Trias period, which immediately succeeds the Carboniferous, the fauna of the Secondary age acquires its definitive character; here the Reptiles first appear. They are huge Crocodilian animals, belonging to a peculiar order, the Rhizodonts, (*Protosaurus*, *Notosaurus*, and *Labyrinthodon.*) The well-known discoveries of Professor Hitchcock, in the red sandstone of the Connecticut, have made us acquainted

* This circumstance, in connection with the absence of Reptiles, has caused the coal-measures to be generally referred to the Palæozoic epoch. But there are other reasons which induce us to unite the carboniferous period with the secondary age, especially when considering that here the land animals first appear, whereas, in the Palæozoic age, there are only marine animals, breathing by gills; and, also, that a luxuriant terrestrial vegetation was developed at that epoch.

with a great number of birds' tracks (Fig. 158, *a*, *b*) belonging to this epoch, for the most part indicating birds of gigantic size. These impressions, which he has designated under the name of *Ornithichnites*, are some of them eighteen inches

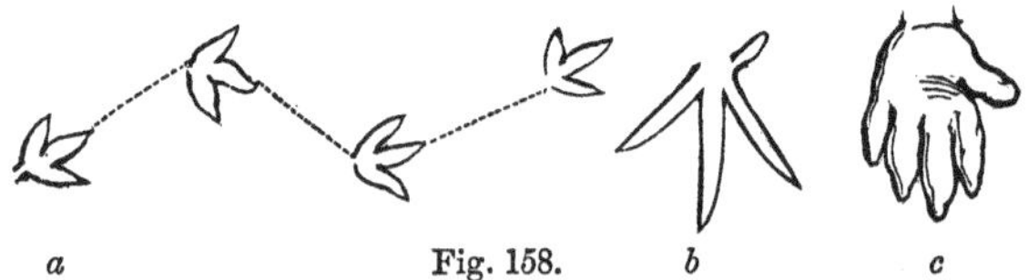
a Fig. 158. *b* *c*

in length, and five feet apart, far exceeding in size the tracks of the largest ostrich. Other tracks, of a very peculiar shape, have been found in the red sandstone of Germany, and in Pennsylvania. They were probably made by Reptiles which have been called *Cheirotherium*, from the resemblance of the track to a hand, (*c*.) The Mollusks, Articulates, and Radiates of this period, approach to the fauna of the succeeding period.

484. The fauna of the Oölitic formation is remarkable for the great number of gigantic Reptiles which it contains. In

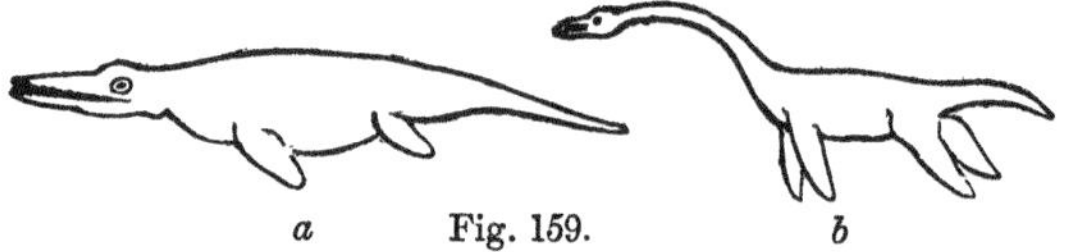
a Fig. 159. *b*

this formation we find those enormous Amphibia, known under the names *Ichthyosaurus*, *Plesiosaurus*, *Megalosaurus*, and *Iguanodon*. The first, in particular, the *Ichthyosaurus*, (Fig. 159, *a*,) greatly abounded on the coast of the continents of that period, and their skeletons are so well preserved, that we are enabled to study even the minutest details of their structure, which differs essentially from that of the Reptiles of the present day. In some respects they form an intermediate link between the Fishes and Mammals, and may be considered as the prototypes of the Whales, having, like them, limbs in the form of oars. The *Plesiosaurus* (*b*) agrees, in many respects, with the *Ichthyosaurus*, in its structure, but is easily distinguished by its long neck, which resembles somewhat the neck of some of our birds. A still more extraordinary Reptile is the *Pterodactylus*, (Fig. 160,) with its long fingers, like those of a bat, and which is thought to have been capable of flying.

Fig. 160.

485. It is also in the upper stages of this formation that we first meet with Tortoises. Here also we find impressions of several families of insects, (*Libellulæ*, *Coleoptera*, *Ichneumons*, *&c.*) Finally, in these same stages, the slates of Stonesfield, the first traces of Mammals are found, namely, the jaws and teeth of animals having some resemblance to the Opossum.

486. The department of Mollusks is largely represented in all its classes. The peculiar forms of the primary age have almost all disappeared, and are replaced by a much greater variety of new forms. Of the Brachiopods only one

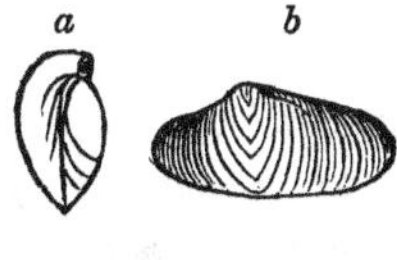

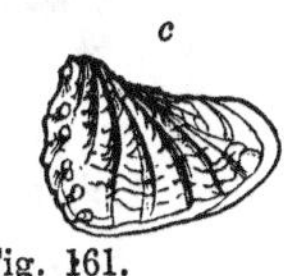

Fig. 161.

type is very abundant, namely, the *Terebratula*, (Fig. 161, *a*.) Among the other Bivalves there are many peculiar forms, as the *Goniomya* (*b*) and the *Trigonia*, (*c*.) The Gasteropods display a great variety of species, and also the Cephalopods, among which the Ammonites are the most prominent, (*d*.) There are also found, for the first time, numerous representatives of the Cuttle-fishes, under the form of *Belemnites*,

(Fig. 162,) an extinct type of animals, protected by a sheath, and terminating in a conical body, somewhat similar to the bone of the *Sepia*, which commonly is the only part preserved, (*b*.)

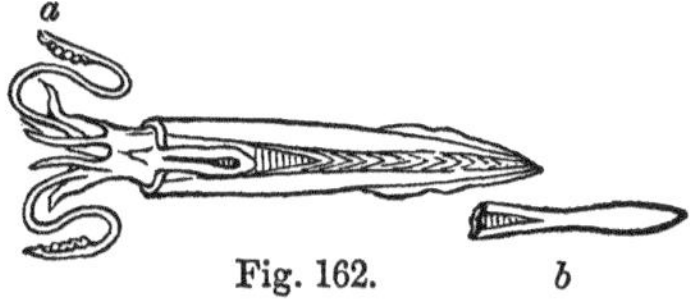

Fig. 162.

487. The variety is not less remarkable among the Radiates. There are to be found representatives of all the classes; even traces of Jelly-fishes have been made out in the slate of Solenhofen, in Bavaria. The Polyps were very abundant at that epoch, especially in the upper stages, one of which has received the name of Coral-rag. Indeed, there are found whole reefs of corals in their natural position, similar to those which are seen in the islands of the

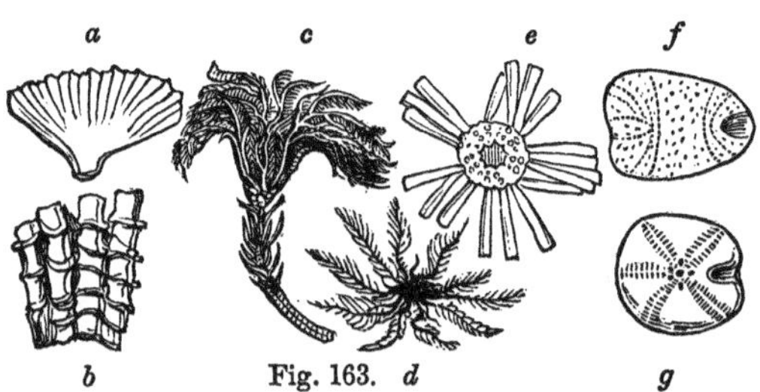

Fig. 163.

Pacific. Among the most remarkable types of stony Polyps may be named the fan-like Lobophyllia, (*L. flabellum*, *a*,) and various forms of tree-corals, *Lithodendron pseudostylina*, *b*.) But the greatest variety exists among the Echinoderms. The Crinoids are not quite so numerous as in former ages. Among the most abundant are the *Pentacrinus*, (*c*.) There are also Comatula-like animals, that is to say, free Crinoids, (*Pterocoma pinnata*, *d*.) Many Star-fishes are likewise observed in the various stages of this formation. Finally, there is an extraordinary variety of

Echini, among them Cidaris, (*e*,) with large spines, and several other types not found before, as, for example, the *Dysaster*, (*f*) and the *Nucleolites*, (*g*.)

488. The fauna of the Cretaceous period bears the same general characters as the Oölitic, but with a more marked tendency towards existing forms. Thus, the *Ichthyosauri* and *Plesiosauri*, that characterize the preceding epoch, are succeeded by gigantic Lizards, more nearly approaching the Reptiles of the present day. Among the Mollusks, a great number of new forms appear, especially among the Cephalopods,* some of which resemble the

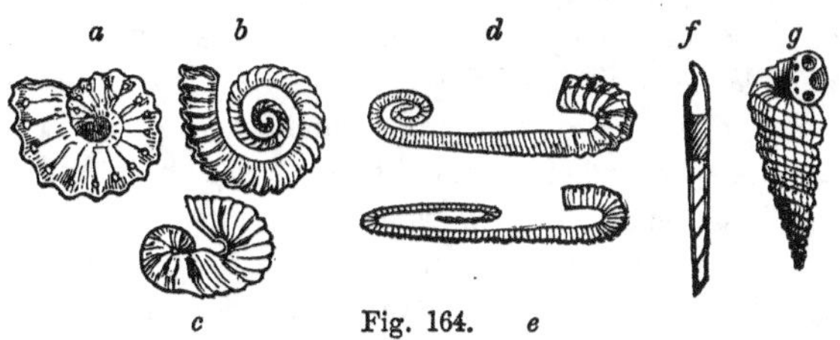

Fig. 164.

Gasteropods in their shape, but are nevertheless chambered. The Ammonites themselves are quite as numerous

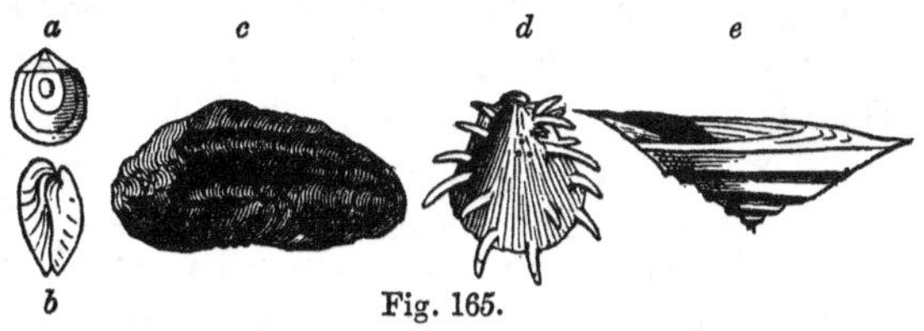

Fig. 165.

as in the Oölitic period, and are in general much ornamented, (*a*.) The Acephala furnish us, also, with peculiar types, not occurring elsewhere, *Magas*, (*a*,) the *Inoceramus*,

* (*a*) *Ammonites;* (*b*) *Crioceras;* (*c*) *Scaphites;* (*d*) *Ancyloceras;* (*e*) *Hamites;* (*f*) *Baculites;* (*g*) *Turrilites.*

(*b*,) the *Hippurites*, (*c*,) and peculiar *Spondyli*, with long spines, (*d*.) There is also a great variety of Gasteropods, among which are some peculiar forms of *Pleu-*

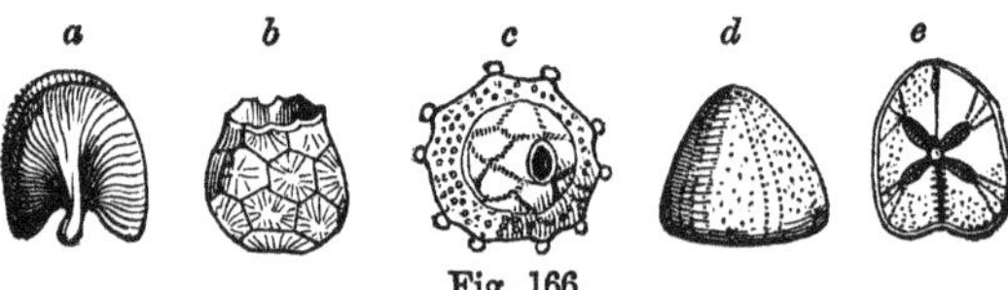

Fig. 166.

rotomaria, (*e*.) The Radiates are not inferior to the others in variety.*

489. TERTIARY AGE. *Reign of Mammals.*—The most significant characteristic of the Tertiary faunas is their great resemblance to those of the present epoch. The animals belong in general to the same families, and mostly to the same genera, differing only as to the species. And the specific differences are sometimes so slightly marked, that a considerable familiarity with the subject is required, in order readily to detect them. Many of the most abundant types of former epochs have now disappeared. The changes are especially striking among the Mollusks, the two great families of Ammonites and Belemnites, which present such an astonishing variety in the Oölitic and Cretaceous epochs, being now completely wanting. Changes of no less importance take place among the Fishes, which are for the most part covered with horny scales, like those of the present epoch, while in earlier ages they were generally covered with enamel. Among the Radiata, we see the family of Crinoids reduced to a very few species, while, on the other hand, a great number of new Star-fishes and Sea-urchins make their appearance. There are, besides, innumerable

* (*a*) *Diploctenium cordatum*; (*b*) *Marsupites*; (*c*) *Salenia*; (*d*) *Galerites*; (*e*) *Micraster cor-anguinum*.

remains of a very peculiar type of animals, almost unknown to the former ages, as well as to the present period. They are little chambered shells, known to geologists under the name of *Nummulites*, from their coin-like appearance, and form very extensive layers of rocks, (Fig. 167.)

Fig. 167.

490. But what is more important in a philosophical point of view is, that aquatic animals are no longer predominant in Creation. The great marine or amphibian reptiles give place to numerous mammals of great size; for which reason, we have called this age the *Reign of Mammals*. Here are also found the first distinct remains of fresh-water animals.

491. The lower stage of this formation is particularly characterized by great Pachyderms, among which we may mention the *Paleotherium* and *Anoplotherium*, which have acquired such celebrity from the researches of Cuvier. These animals, among others, abound in the Tertiary formations of the neighborhood of Paris. The Paleotheriums, of

Fig. 168. Fig. 169.

which several species are known, are the most common; they resemble, (Fig. 168,) in some respects, the Tapirs, while the Anoplotheriums are more slender animals, (Fig. 169.) On this continent are found the remains of a most extraordinary animal of gigantic size, the Basilosaurus, a true cetacean. Finally, in these stages, the earliest remains of Monkeys have been detected.

492. The fauna of the upper stage of the Tertiary formation approaches yet more nearly to that of the present epoch. Besides the Pachyderms, that were also predominant in the lower stage, we find numbers of carnivorous animals, some of them much surpassing in size the lions and tigers of our day. We meet also gigantic Edentata, and Rodents of great size.

493. The distribution of the Tertiary fossils also reveals to us the important fact, that, in this epoch, animals of the same species were circumscribed in much narrower limits than before. The earth's surface, highly diversified by mountains and valleys, was divided into numerous basins, which, like the Gulf of Mexico, or the Mediterranean of this day, contained species not found elsewhere. Such was the basin of Paris, that of London, and, on this continent, that of South Carolina.

494. In this limitation of certain types within certain bounds, we distinctly observe another approach to the present condition of things, in the fact that groups of animals which occur only in particular regions are found to have already existed in the same regions during the Tertiary epoch. Thus the Edentata are the predominant animals in the fossil fauna of Brazil as well as in its present fauna; and Marsupials were formerly as numerous in New Holland as they now are, though in general of much larger size.

495. The Modern Epoch. *Reign of Man.*—The Present epoch succeeds to, but is not a continuation of, the Tertiary age. These two epochs are separated by a great geological event, traces of which we see every where around us The climate of the northern hemisphere, which had been, during the Tertiary epoch, considerably warmer than now, so as to allow of the growth of palm-trees in the temperate zone of our time, became much colder at the end of this period, causing the polar glaciers to advance south, much beyond

their previous limits. It was this ice, either floating like icebergs, or, as there is still more reason to believe, moving along the ground, like the glaciers of the present day, that, in its movement towards the South, rounded and polished the hardest rocks, and deposited the numerous detached fragments brought from distant localities, which we find every where scattered about upon the soil, and which are known under the name of *erratics*, *boulders*, or *grayheads*. This phase of the earth's history has been called, by geologists, the *Glacial* or *Drift period*.

496. After the ice that carried the erratics had melted away, the surface of North America and the North of Europe was covered by the sea, in consequence of the general subsidence of the continents. It is not until this period that we find, in the deposits known as the diluvial or pleistocene formation, incontestable traces of the species of animals now living.

497. It seems, from the latest researches of Geologists, that the animals belonging to this period are exclusively marine; for, as the northern part of both continents was covered to a great depth with water, and only the summits of the mountains were elevated above it, as islands, there was no place in our latitudes where land or fresh-water animals could exist. They appeared therefore at a later period, after the water had again retreated; and as, from the nature of their organization, it is impossible that they should have migrated from other countries, we must conclude that they were created at a more recent period than our marine animals.

498. Among these land animals which then made their appearance, there were representatives of all the genera and species now living around us, and besides these, many types now extinct, some of them of a gigantic size, such as the Mastodon, the remains of which are found in the upper-

most strata of the earth's surface, and probably the very last large animal which became extinct before the creation of man.*

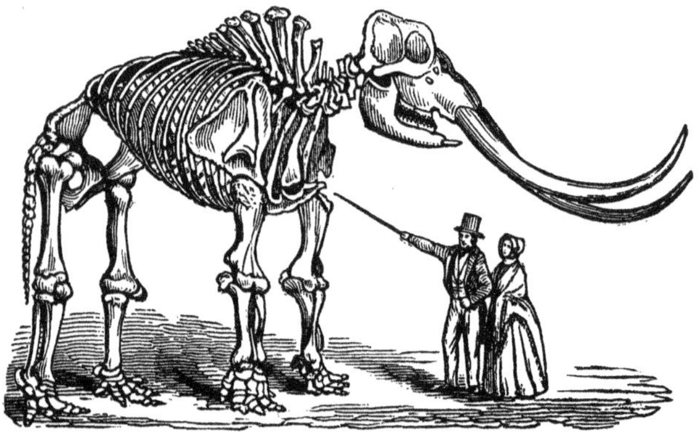

Fig. 170.

499. It is necessary, therefore, to distinguish two periods in the history of the animals now living; one in which the marine animals were created, and a second, during which the land and fresh-water animals made their appearance, and at their head MAN.†

CONCLUSIONS.

500. From the above sketch it is evident that there is a manifest progress in the succession of beings on the surface

* The above diagram is a likeness of the splendid specimen disinterred at Newburg, N. Y., now in the possession of Dr. J. C. Warren, in Boston; the most complete skeleton which has ever been discovered. It stands nearly twelve feet in height, the tusks are fourteen feet in length, and nearly every bone is present, in a state of preservation truly wonderful.

† The former of these phases is indicated in the frontispiece, by a narrow circle, inserted between the upper stage of the Tertiary formation and the *Reign of Man* properly so called.

of the earth. This progress consists in an increasing similarity to the living fauna, and among the Vertebrates, especially, in their increasing resemblance to Man.

501. But this connection is not the consequence of a direct lineage between the faunas of different ages. There is nothing like parental descent connecting them. The Fishes of the Palæozoic age are in no respect the ancestors of the Reptiles of the Secondary age, nor does Man descend from the Mammals which preceded him in the Tertiary age. The link by which they are connected is of a higher and immaterial nature ; and their connection is to be sought in the view of the Creator himself, whose aim, in forming the earth, in allowing it to undergo the successive changes which Geology has pointed out, and in creating successively all the different types of animals which have passed away, was to introduce Man upon the surface of our globe. Man is the end towards which all the animal creation has tended, from the first appearance of the first Palæozoic Fishes.

502. In the beginning His plan was formed, and from it He has never swerved in any particular. The same Being who, in view of man's moral wants, provided and declared, thousands of years in advance, that "the seed of the woman shall bruise the serpent's head," laid up also for him in the bowels of the earth those vast stores of granite, marble, coal, salt, and the various metals, the products of its several revolutions; and thus was an inexhaustible provision made for his necessities, and for the development of his genius, ages in anticipation of his appearance.

503. To study, in this view, the succession of animals in time, and their distribution in space, is, therefore, to become acquainted with the ideas of God himself. Now, if the succession of created beings on the surface of the globe is the realization of an infinitely wise plan, it follows that there

must be a necessary relation between the races of animals and the epoch at which they appear. It is necessary, therefore, in order to comprehend Creation, that we combine the study of extinct species with that of those now living, since one is the natural complement of the other. A system of Zoölogy will consequently be true, in proportion as it corresponds with the order of succession among animals.

INDEX AND GLOSSARY.

LIST OF THE MOST IMPORTANT AUTHORS

WHO MAY BE CONSULTED IN REFERENCE TO THE SUBJECTS TREATED IN THIS WORK.

GENERAL ZOÖLOGY.

Aristotle's Zoölogy; Linnæus, System of Nature; Cuvier's Animal Kingdom; Oken's Zoölogy; Humboldt's Cosmos, and Views of Nature; Spix, History of Zoölogical Systems; Cuvier's History of the Natural Sciences.

ANATOMY AND PHYSIOLOGY.

Henle's General Anatomy; and most of the larger works on Comparative Anatomy, Physiology, and Botany, such as those of Hunter, Cuvier, Meckel, Müller, Todd and Bowman, Grant, Owen, Carpenter, Rymer Jones, Hassall, Quain and Sharpey, Bourgery and Jacob, Wagner, Siebold, Milne Edwards, Carus, Schleiden, Burmeister, Lindley, Robert Brown, Dutrochet, Decandolle, A. Gray.

ON SPECIAL SUBJECTS OF ANATOMY AND PHYSIOLOGY MAY BE CONSULTED

Schwann, on the Conformity in the Structure and Growth of Animals and Plants.

Dumas and Boussingault, on Respiration in Animals and Plants.

Valentin, on Tissues; and Microscopic Anatomy of the Senses.

Sœmmering, Figures of the Eye and Ear.

Kölliker, Theory of the Animal Cell.

Breschet, on the Structure of the Skin.

Locomotion; Weber, and Dugès.

Teeth; Fred. Cuvier, Geoff. St. Hilaire, Owen, Nasmyth, Retzius.

Blood; Döllinger, Barry.

Digestion; Spallanzani, Valentin and Brunner, Dumas and Boussin gault, Liebig, Matteucci, Beaumont.

INSTINCT AND INTELLIGENCE.

Kirby, Blumenbach, Spurzheim, Combe.

EMBRYOLOGY.

D'Alton, Von Baer, Purkinje, Wagner, Wolfe, Rathke, Bischoff, Velpeau, Flourens, Barry, Leidy.

PECULIAR MODES OF REPRODUCTION.

Ehrenberg, Trembly, Rösel, Sars, Lovèn, Steenstrup, Van Beneden.

METAMORPHOSIS.

St. Merian, Rösel, De Geer, Harris, Kirby and Spence, Burmeister, Reaumur.

GEOGRAPHICAL DISTRIBUTION.

Zimmerman, Milne Edwards, Swainson, A. Wagner, Forbes, Pennant, Richardson, Ritter, Guyot.

GEOLOGY.

The Works of Murchison, Phillips, Lyell, Mantell, Hugh Miller, Agassiz, D'Archiac, De Beaumont, D'Orbigny, De Verneuil, Cuvier, Brongniart, Deshayes, Morton, Hall, Conrad, Hitchcock, Troost, and the Reports on the various local Geological Surveys.

Very many of the papers of the authors above referred to are not published in separate volumes, but are scattered through the volumes of Scientific Periodicals; such as the

Transactions of the Royal Society of London.
Annals and Magazine of Natural History.
Annales, and Archives, du Museum d' Hist. Naturelle.
Annales des Sciences Naturelles.
Wiegmann's Archiv für Naturgeschichte.
Müller's Archiv.
Oken's Isis.
Berlin Transactions.
Transactions of the American Philosophical Society.
Memoirs of the American Academy.
Journal of the Academy of Nat. Sciences, Philadelphia.
Silliman's Journal.
Journal of Boston Society of Natural History.

END OF PART I.

www.ingramcontent.com/pod-product-compliance
Lightning Source LLC
LaVergne TN
LVHW020056110826
845151LV00005B/1489
9781425522063